EPIOPTICS-8

THE SCIENCE AND CULTURE SERIES — PHYSICS

1. Perspectives for New Detectors in Future Supercolliders, 1991
2. Data Structures for Particle Physics Experiments: Evolution or Revolution?, 1991
3. Image Processing for Future High-Energy Physics Detectors, 1992
4. GaAs Detectors and Electronics for High-Energy Physics, 1992
5. Supercolliders and Superdetectors, 1993
6. Properties of SUSY Particles, 1993
7. From Superstrings to Supergravity, 1994
8. Probing the Nuclear Paradigm with Heavy Ion Reactions, 1994
9. Quantum-Like Models and Coherent Effects, 1995
10. Quantum Gravity, 1996
11. Crystalline Beams and Related Issues, 1996
12. The Spin Structure of the Nucleon, 1997
13. Hadron Colliders at the Highest Energy and Luminosity, 1998
14. Universality Features in Multihadron Production and the Leading Effect, 1998
15. Exotic Nuclei, 1998
16. Spin in Gravity: Is It Possible to Give an Experimental Basis to Torsion?, 1998
17. New Detectors, 1999
18. Classical and Quantum Nonlocality, 2000
19. Silicides: Fundamentals and Applications, 2000
20. Superconducting Materials for High Energy Colliders, 2001
21. Deep Inelastic Scattering, 2001
22. Electromagnetic Probes of Fundamental Physics, 2003
23. Epioptics-7, 2004
24. Symmetries in Nuclear Structure, 2004
25. Innovative Detectors for Supercolliders, 2003
26. Complexity, Metastability and Nonextensivity, 2004
27. Epioptics-8, 2004

THE SCIENCE AND CULTURE SERIES — PHYSICS

EPIOPTICS-8

Proceedings of the 33rd Course of the International School of Solid State Physics

Erice, Italy 20 – 26 July 2004

Editor

Antonio Cricenti

Series Editor

A. Zichichi

World Scientific

NEW JERSEY • LONDON • SINGAPORE • BEIJING • SHANGHAI • HONG KONG • TAIPEI • CHENNAI

Published by

World Scientific Publishing Co. Pte. Ltd.

5 Toh Tuck Link, Singapore 596224

USA office: 27 Warren Street, Suite 401-402, Hackensack, NJ 07601

UK office: 57 Shelton Street, Covent Garden, London WC2H 9HE

British Library Cataloguing-in-Publication Data
A catalogue record for this book is available from the British Library.

EPIOPTICS-8

ISBN 981-256-743-7

Printed in Singapore by World Scientific Printers (S) Pte Ltd

PREFACE

This special Volume of World Scientific contains the Proceedings of the 8th Epioptics Workshop, held in the Ettore Majorana Foundation and Centre for Scientific Culture, Erice, Sicily, from July 20 to 26, 2004. The Workshop was the 8th in the Epioptics series and the 33rd of the International School of Solid State Physics. Antonio Cricenti from CNR Istituto di Struttura della Materia and Theo Rasing from the University of Njimegen, were the Directors of the Workshop. The Advisory Committee of the Workshop included Y. Borensztein from U. Paris VII (F), R. Del Sole from U. Roma II Tor Vergata (I), D. Aspnes from NCSU (USA), O. Hunderi from U. Trondheim (N), J. McGilp from Trinity College Dublin (Eire), W. Richter from TU Berlin (D), N. Tolk from Vanderbilt University (USA), and P. Weightman from University Liverpool (UK). Forty-two scientists from 13 countries attended the Workshop.

The Workshop has brought together researchers from universities and research institutes who work in the fields of (semiconductor) surface science, epitaxial growth, materials deposition and optical diagnostics relevant to (semiconductor) materials and structures of interest for present and anticipated (spin) electronic devices. The Workshop was aimed at assessing the capabilities of state-of-the-art optical techniques in elucidating the fundamental electronic and structural properties of semiconductor and metal surfaces, interfaces, thin layers, and layer structures, and assessing the usefulness of these techniques for optimization of high quality multilayer samples through feedback control during materials growth and processing. Particular emphasis is dedicated to theory of non linear optics and to dynamical processes through the use of pump-probe techniques together with the search for new optical sources. Some new applications of Scanning Probe Microscopy to Material science and biological samples, dried and *in vivo*, with the use of different laser sources have also been presented. Materials of particular interest have been silicon, semiconductor-metal interfaces, semiconductor and magnetic multilayers and III-V compound semiconductors. The Workshop as well as the notes collected in this Volume combined the tutorials aspects adequate to a School with some of the most advanced topics in the field, which better characterize the Workshop.

We wish to thank our sponsors, the Italian National Research Council (CNR) and the Sicilian Regional Government for facilitating a most successful Workshop. We are grateful to the Director of the International School of Solid State Physics,

Prof. G. Benedek, to the Director of the Ettore Majorana Centre, Prof. A. Zichichi, and to the Centre staff members for the excellent support, organization and hospitality provided.

Antonio Cricenti

CONTENTS

AB-INITIO THEORIES FOR THE CALCULATION OF ELECTRONIC EXCITED STATES PROPERTIES

O. PULCI[1], M. MARSILI[1], E. LUPPI[2], C. HOGAN[1], E. DEGOLI[3]
AND R. DEL SOLE[1]

[1] *European Theoretical Spectroscopy Facility (ETSF) and CNR-INFM, Dept. of Physics, University of Rome "Tor Vergata", Via della Ricerca Scientifica 1, I-00133 Roma, Italy*

[2] *CNR-INFM-S3, Dept. of Physics, University of Modena and Reggio Emilia, Via Campi 213 A, I-41100 Modena, Italy*

[3] *CNR-INFM-S3, Dip. di Scienze e Metodi dell'Ingegneria, Universita' di Modena e Reggio Emilia, via Fogliani 1, I-42100 Reggio Emilia, Italy*

We review the state of the art of the theoretical approaches for *ab initio* studies of the electronic and optical properties of matter. Examples within Density Functional Theory, Many-Body perturbation Theory and Time Dependent Density Functional Theory are presented and discussed. The static and the time-dependent DFT avoid dealing directly with the Many-Body equation by mapping the interacting system into a fictitious non-interacting system (which is then described by single-particle equations); the Green's function approach, instead, maps the Many-Body electronic problem to a system of quasi-particles, which describe the excitations in terms of a particle of finite life time, that represent the extra electron (and/or the extra hole added to the system) plus its screened interaction with the electrons of the system. Advantages and drawbacks of the different approaches are discussed.

1. Introduction

Structural, electronic and optical properties of complex systems are nowadays accessible thanks to the impressive development of theoretical approaches and of computer power. Surfaces, nanostructures, and even biological systems can now be studied within ab-initio methods. In principle within the Born-Oppenheimer approximation to decouple the ionic and electronic dynamic, the equation that governs the physics of all those systems is the Many-Body equation:

$$(\sum_i -\frac{1}{2}\nabla^2_{\mathbf{r}_j} + V_{ext} + \frac{1}{2}\sum_{i\neq j}\frac{1}{|\mathbf{r}_i - \mathbf{r}_j|})\Psi(r_1, r_2, ..r_N) = E\Psi(r_1, r_2, ..r_N) \quad (1)$$

with V_{ext} ionic potential.

A direct solution of equation (1) is a formidable task for realistic systems. It is hence necessary to resort to approximations. We review in this paper the theoretical approaches used: Density Functional Theory (DFT) for ground state properties, GW for band structure calculations (charged excitations), and the Bethe Salpeter approach and/or the Time Dependent DFT for optical spectra (neutral excitations).

2. Density Functional theory: ground state properties

Based on the seminal work of Hohenberg and Kohn in 1964[2], Density Functional Theory (DFT) proved to be a formidable tool to get ground state properties also of very complex systems (for an authoritative review of DFT, see for example [3]). The power of this theory lies in the fact that its fundamental quantity is not anymore the extremely complex N-variables dependent ground state wave function but the *simpler* one-variable dependent electronic density. In fact in their work, Hohenberg and Kohn proved that all ground state properties of an interacting electronic system, including the many-body wave function and total energy E, could be expressed in terms of unique functionals of the electronic density alone. Furthermore, the variational principle ensures that for a given density $n(\mathbf{r})$, the total energy functional $E\left[n(\mathbf{r})\right] \geq E_{GS}$ (the ground state energy of the system), and in fact finds its minimum at the ground state density n_{GS}, for which $E\left[n_{GS}\right] = E_{GS}$. The explicit form of the density functional is not known, and may not even exist. A real breakthrough came a year later with the work of Kohn and Sham [4]. By separating out the terms contributing to the total energy E, Kohn and Sham proposed a single particle scheme for obtaining the ground state density and total energy. In this way, the more complicated interacting system is mapped onto a simpler non-interacting (albeit fictitious) system with Hamiltonian

$$[-\frac{1}{2}\nabla^2 + v_{ext} + v_H + V_{xc}]\phi_i(\mathbf{r}) = \varepsilon_i\phi_i(\mathbf{r}) \tag{2}$$

which is constrained to have the same ground state density as the interacting system:

$$n(r) = \sum_i f_i|\phi_i(\vec{r})|^2 \tag{3}$$

(here f_i is the occupation number of state i). In Eq. 2, v_H is the Hartree potential and $V_{xc} = \delta E_{xc}\left[n\right]/\delta n$ is the exchange and correlation potential of the interacting system. Both depend self-consistently on the electronic density. $E_{xc}\left[n\right]$ contains the exchange and correlation part of the total

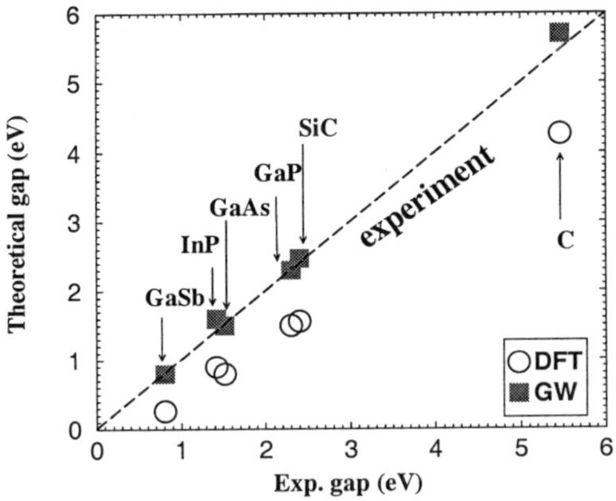

Figure 1. Calculated [6] electronic gaps for several semiconductors, compared with the experimental values [7]. circles: DFT-LDA calculations; squares: GW calculations. The straight line represents the experimental gaps.

energy not included in the Hartree term plus the difference between the kinetic energy of the fictitious non interacting system and of the real one.

Given an approximation for $E_{xc}[n]$, it is hence possible to solve the Kohn-Sham equations (2,3) self consistently and directly obtain the density of the real interacting system. Once the density is known, the ground state energy of the interacting system can be determined. Minimization with respect to the ionic degree of freedoms allows also to determine the ground state geometry.

The simplest and most common approximation to E_{xc} is the Local Density Approximation (LDA) given by[5]:

$$E_{xc}^{LDA} = \int d\mathbf{r} n(\mathbf{r}) \epsilon_{xc}^{\text{heg}}(n), \tag{4}$$

where $\epsilon_{xc}^{\text{heg}}(n)$ is the exchange-correlation energy (per electron) of a homogeneous electron gas of density n. The eigenvalues of the Kohn and Sham equations are often interpreted as one electron excitation energies. In this way, DFT can be used to calculate band structures: although remarkable qualitative agreement with experiment is often observed, the electronic gaps of semiconductors are systematically underestimated. As seen in Fig. 1, the theoretical DFT results (circles) for the electronic gaps always lie below the straight line representing the experimental values.

3. The GW approach: study of charged excitations

Underestimation of the electronic gaps in DFT calculations finds its origin not in a deficiency of the theory, but in the uncorrect use we make of the Kohn-Sham equations. In fact, DFT is an exact theory for ground state properties, but there is no strict theoretical justification to use it to obtain excitation energies.

Thinking that when we look at the experimental band structure we are actually looking at the results of experiments (direct and inverse photoemission) that probe the excitation energies of the system upon removal or addition of an electron, we see that it is straightforward to pass to the Green's function formalism. In fact, after a Fourier transform into the frequency domain, the single particle Green's function, defined as probability amplitude of an electron (hole) propagation from $\mathbf{r}_1$ to $\mathbf{r}_2$ in the time interval $(t_2 - t_1)$, has poles at the energy differences between the N electrons system ground state and the N $\pm$ 1 electrons excited state. In the Fourier space the Green's function can be formally written as

$$G(\mathbf{r}_1, \mathbf{r}_2, \omega) = \sum_s \frac{\Psi_{0s}(\mathbf{r}_1)\Psi^+_{s0}(\mathbf{r}_2)}{\omega - \epsilon_s - \mu + i\delta} + \sum_{s'} \frac{\Psi^+_{0s'}(\mathbf{r}_2)\Psi_{s'0}(\mathbf{r}_1)}{\omega + \epsilon_{s'} - \mu - i\delta} \tag{5}$$

where s and s' run over the N+1 and N-1 electrons excited states, $\Psi_{0s}(\mathbf{r})$ $(\Psi_{0s'}(\mathbf{r}))$ is the expectation value of the creation operator of an electron at position $\mathbf{r}$ between the N particle ground state and the $N+1(N-1)$ particle excited state labelled by $s(s\prime)$, and $\epsilon_s(\epsilon_{s\prime})$ is the energy of the N+1 (N-1) excited states; the infinitesimally small imaginary term $i\delta$ is needed for the convergence of the Fourier transform over the time variable.

So, in principle if the single particle Green's function of a the system is known, the determination of the positions of its poles allow the access to the single particle excitation energies.

The Green's function can not be calculated exactly for realistic systems. It obeys a Dyson equation of the form

$$G(1,2) = G_0(1,2) + \int d(34) G_0(1,3)\Sigma(3,4)G(4,2) \tag{6}$$

where G_0 is the non-interacting Green's function and Σ is the self-energy, a non-hermitian, non-local and energy dependent operator.
In this framework, by introducing the equation of motion for G, we can obtain the so called 'quasi-particle equation':

$$\left[-\frac{\nabla^2}{2} + V_{ext}(\mathbf{r}_1) + V_H(\mathbf{r}_1)\right] \phi_{sk}(\mathbf{r}_1, \omega) +$$

$$\int d\mathbf{r}_2 \Sigma(\mathbf{r}_1, \mathbf{r}_2, \omega) \phi_{sk}(\mathbf{r}_2, \omega) = \epsilon_{sk}^{QP}(\omega) \phi_{sk}(\mathbf{r}_1, \omega) \tag{7}$$

where ϵ_{sk}^{QP} are the poles of the Green's function, so that the solution of eq. (7) gives directly the electronic band structure of the system. It is worth noticing that this equation reduces to the Hartree equations when Σ=0, to the Hartree-Fock ones when $\Sigma = iGV$, and to the Kohn-Sham equations when, instead of Σ, a local, hermitian and energy independent operator is taken: $\Sigma = V_{xc}^{KS}(r)$.
Now, *a-posteriori*, we can understand why the Kohn and Sham eigenvalues often produce a band structure in qualitative agreement with experiments: somehow $V_{xc}^{KS}(r)$ provides a quite good approximation for the *true* self-energy Σ.

We see from eq.(7) that in order to get G we need a suitable expression for Σ. However, unlike the case of DFT, theory provides an exact set of closed equations,the Hedin equations[9], which together with eq.(6) define implicitly the self-energy Σ. Neglecting the vertex corrections and iterating Hedin's equation the self-energy operator takes the form $\Sigma = iGW$, which is the famous 'GW' approximation [10,11]. Computing quasi-particle energies within the GW approximation, is the state of the art method for band structure calculations; examples of experimental and theoretical results obtained within the GW approximation are shown in Fig.(1): the theoretical GW gaps (squares) are in very good agreement with the experimental ones (straight line) and the problem of underestimation of the electronic gaps, typical of DFT calculations (circles), is overcome.

The GW method can be applied to the calculation of the band structures of solids and surfaces and to the determination of the energy levels of molecules and of atoms. As an example, we show the results obtained for the diamond (111) 2x1 surface. All DFT *ab-initio* calculations show that this surface appears metallic while experimental data show that its electronic structure has a gap of at least 0.5 eV [12]. Hence, we have calculated quasi-particle corrections to the DFT energies, by evaluating the diagonal elements of $(\Sigma - V_{xc}^{DFT})$ between DFT states [13]. The results are shown in Fig.2: a minimum gap of 0.8 eV opens between the surface states at J, leading to a semiconducting surface, and a good agreement with the available photoemission experiments is found.

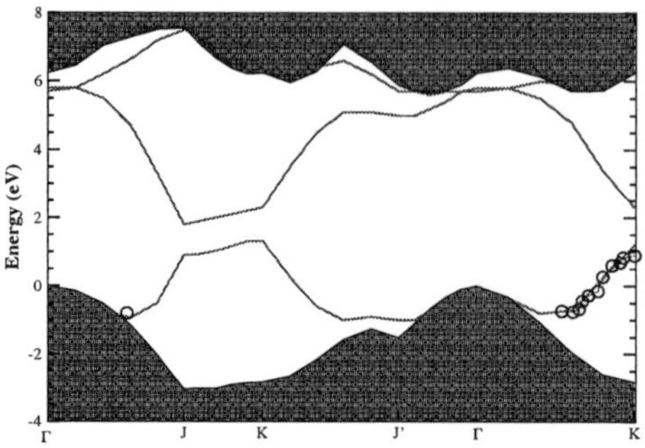

Gap	J	K	Γ	J'
DFT	0.3	0.2	4.5	5.0
GW	0.8	0.9	5.7	6.3

Figure 2. C(111)2x1 band structure calculated within the GW scheme. Crosses: experimental results from [14,12]. The numerical values of the gaps between surface states within DFT and GW are listed. All energies are expressed in eV.

4. Optical properties: study of neutral excitations

Whereas in photoemission the final state of the system is charged, since one electron has been removed or added to it, in optical absorption experiments instead, the system is left in what is called a neutral excited state. Excitons are examples of neutral excited states that are commonly detected by optical spectroscopy; in a very schematic way excitons can be thought as bounded electron-hole states [16].

The physical quantity directly connected to optical spectra is the macroscopic dielectric function ϵ_M. This is related to the microscopic dielectric function $\epsilon_{\mathbf{G},\mathbf{G}'}(\mathbf{q},\omega)$ by a macroscopic limit which, following the derivation of Adler and Wiser [15], is given by:

$$\epsilon_M(\omega) = \lim_{\mathbf{q}\to 0} \frac{1}{(\epsilon(q,\omega))^{-1}_{\mathbf{G}=0,\mathbf{G}'=0}}. \tag{8}$$

The inverse of the microscopic dielectric matrix can be computed in terms of the reducible polarizability χ knowing that $\epsilon^{-1} = 1 + v\chi$, where v is the bare Coulomb interaction. In turn, the reducible polarizability satisfies a Dyson-like equation that connects it to the irreducible polarizability P [8]:

$$\chi = P + Pv\chi \tag{9}$$

P can be computed within the Hedin's equations scheme [9] at different levels of approximation. If we neglect the vertex corrections, that corresponds in neglecting the interaction between the holes and the electron formed during the polarization process, we obtain the so called random phase approximation (RPA) for the dielectric matrix. Within this approximation the irreducible polarizability is given by:

$$P_{IQP}(x,x';\omega) = 2\sum_{i,j}(f_i - f_j)\frac{\phi_i(x)\phi_j^*(x)\phi_i^*(x')\phi_j(x')}{\epsilon_i - \epsilon_j - \omega - i\eta} \tag{10}$$

here (i,j) are single quasi-particle state labels, f_i and ϕ_i are the corresponding occupation and wave function. From this equation, neglecting local field effects, the response to a longitudinal field, for $q \to 0$, is:

$$\epsilon_M(\omega) = 1 - \lim_{q\to 0} v(q)\sum_{vc}\frac{\left|\langle v|\exp^{-iqr}|c\rangle\right|^2}{\omega - (\epsilon_c - \epsilon_v) - i\eta} \tag{11}$$

$|v\rangle$ ($|c\rangle$) represent valence (conduction) single particle state and ϵ_v (ϵ_c) is its corresponding energy. The spectra, within RPA, (eq. 11) is build up by the sum of all possible independent transitions of an electron from the valence to the conduction states and the structures are located at the energy required for each transition. This is not surprising since we have neglected the vertex correction and thus, as already mentioned, we are dealing with an independent particle scheme. Within this framework it is in fact natural to think that each possible transition contributes independently from the others to the spectra.

Comparing the RPA spectra with experiment, usually both the peak positions and the line shape are wrong. In particular, depending on which single particle scheme is used, whether DFT or GW, the position of the peaks will be shifted toward lower or higher energy. Despite these shortcomings at the RPA level, a lot of success can nevertheless be achieved with the application of an *a posteriori* rigid shift to the DFT conduction bands (the so-called "scissors" operator), while calculating optical spectra. This approach can give very reasonable results for many semiconducting and insulating systems: first of all, DFT-LDA wave-functions are often good approximations to the GW ones [17], and hence produce reasonable oscillator strengths; secondly, if properly (e.g. empirically) chosen, the applied shift can account for both self-energy and higher-order many body corrections together, albeit in a very approximate way. As an example of this method, we show in Figure 3 the results of calculations at the DFT-LDA/RPA level of the reflectance anisotropy (RA) spectra for two reconstructions of the GaAs(001) surface. Being difference spectra, and hence susceptible to error cancellation, RA spectra are intrinsically well suited to the scissors operator scheme. As can be seen in the figure, excellent agreement is found for the $c(4 \times 4)$ reconstruction, both in the line shape and amplitude; agreement is

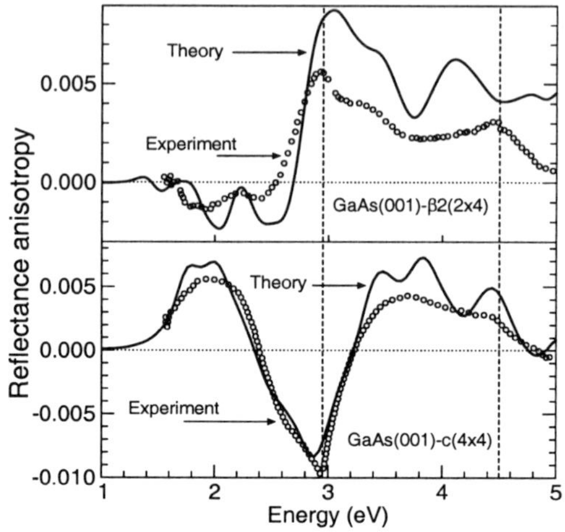

Figure 3. Surface reflectance anisotropy spectra within RPA+shift of GaAs(001) surfaces (solid lines), compared with experimental data (dots)[18]. Top: $\beta2(2\times4)$ reconstruction. Bottom: $c(4\times4)$ mixed dimer reconstruction. Rigid shifts of 0.3 eV and 0.5 eV, respectively, have been used.

slightly worse for the $\beta2(2\times4)$ case, although the main experimental features are still present. However, different energy shifts were needed in each case: 0.5 eV for the $c(4\times4)$, and 0.3 eV for the $\beta2(2\times4)$ (for comparison, the GW correction in bulk GaAs is about 0.8 eV). As a means of analyzing experimental surface spectra of computationally demanding systems, the RPA+scissors approach is therefore quite useful, especially if many-body effects on the spectral line shape are weak; however, if predictive, or quantitatively accurate spectra are required, more theoretically sound approaches are required, as described in the next sections.

4.1. *Bethe-Salpeter equation and excitonic effects*

In some systems, vertex corrections can not be completely neglected if we want to obtain optical spectra in agreement with experiments[19]. Neglecting vertex corrections means to neglect the interaction between the hole and the electron that are formed within the polarization processes of the system. In the description of absorption experiments, in which the excited electron remains in the sample, it is evident that the effects of the electron-hole interaction, i.e. the excitonic effects, can not be neglected. In this paragraph we will see how the effects of the electron-hole interaction in the spectra can be included in MBPT through an effective two body Hamiltonian, the

so called excitonic Hamiltonian.
To include vertex corrections we can think of iterating once more the self-consistency cycle of the Hedin equations. In this way we obtain the so called 'Bethe Salpeter' equation

$$^4\bar{P} =^4 P_{IQP} +^4 P_{IQP} K^4 \bar{P}. \tag{12}$$

for a modified reducible polarizability $\bar{P}$. The kernel K of this integral equation contains an electron-hole exchange $\bar{v}$ and an electron hole attraction $-W$ term:

$$K(1234) = \delta(12)\delta(34)\bar{v}(13) - \delta(13)\delta(24)W(12). \tag{13}$$

It can be shown (see for instance appendix B of Ref. [1]) that the macroscopic dielectric function can be rewritten in terms of this modified reducible polarizability $\bar{P}$:

$$\epsilon_M(\omega) = 1 - \lim_{\mathbf{q}\to 0} v(\mathbf{q})_0 \bar{P}_{\mathbf{G}=\mathbf{G}'=0}(q,\omega). \tag{14}$$

By projecting the Bethe-Salpeter equation (12) in the transition space, i.e. into a basis made of couples of single quasi-particle states, usually labelled by band and wave vector indices, we obtain a two particle Hamiltonian, the so called excitonic Hamiltonian. The excitonic Hamiltonian is not necessarily hermitian, however in calculations just its hermitian part is usually taken into account, which is:

$$H^{2p,exc}_{(vck)(v'c'k')} = (E_{ck} - E_{vk'})\delta_{vv'}\delta_{cc'}\delta_{kk'} + 2v^{v'c'k'}_{vck} - W^{v'c'k'}_{vck} \tag{15}$$

Here E_{ck}, $(E_{vk'})$ are the quasi-particle energies, calculated within the GW approximation, of the states (ck) and (vk'). In terms of the eigenvalues and eigenvectors of the excitonic Hamiltonian, namely:

$$H^{2p,exc}_{(n_1n_2)(n_3n_4)} A^{(n_3n_4)}_{\lambda} = E^{exc}_{\lambda} A^{(n_1n_2)}_{\lambda} \tag{16}$$

the macroscopic dielectric function is:

$$\epsilon_M(\omega) = 1 - \lim_{q\to 0} v(q) \sum_{\lambda} \frac{\left|\sum_{(n_1n_2)} \langle n_1| \exp^{-iqr} |n_2\rangle A^{(n_1n_2)}_{\lambda}\right|^2}{E^{exc}_{\lambda} - \omega - i\eta} \tag{17}$$

where each label n_i represents a couple of band and wave vector indices. If we now compare this formula with the RPA dielectric function (eq.11), we can understand why the inclusion of the excitonic effects influences so strongly the spectra: first of all the poles, hence the positions of the structures, are now located at energies equal to the eigenvalues of the excitonic hamiltonian, and this usually provides a red shift of the spectra; secondly, the line shape is not determined anymore by the independent contribution of each transition, they are instead mixed and each transition is weighted by $A_{\lambda}^{n_1 n_2}$, i.e. by the amount in which it contributes to the excitonic eigenstate labelled by λ.

As an example of an excitonic calculation we show the spectra of bulk LiF. The Brillouin Zone has been sampled with 256 k-points. In the calculation of the macroscopic dielectric functions the contributions of the transitions coming from the two valence and six conduction bands around the gap are included. LiF is a prototype system for bound excitons since it presents a strong excitonic peak at 12.9 eV, with an excitonic binding energy of -1.4 eV . We show in Fig.4 the absorption spectrum calculated within the independent single particle approaches (DFT-RPA and GW-RPA): the DFT calculation gives, as usual, a red shifted spectrum when compared to the experiment. The GW spectrum, obtained within a single particle scheme using GW eigenenergies, shifts the spectrum towards high energies without improving the agreement. The Bethe Salpeter approach, instead, well reproduces the bound exciton below gap and also the spectrum at higher energies, thus confirming to be the state of the art technique for a proper description of optical properties with the inclusion of excitonic effects.

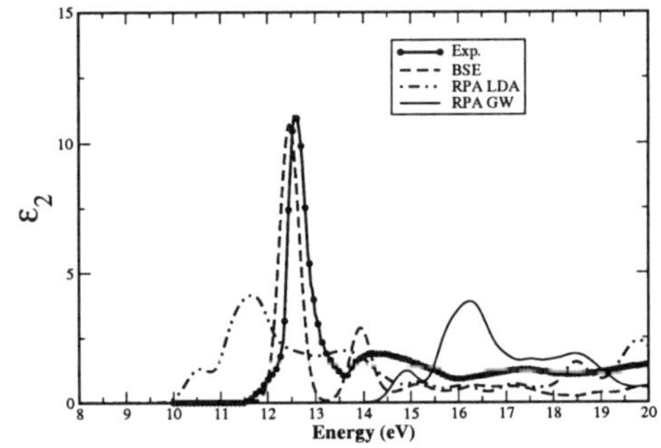

Figure 4. LiF optical spectra. Dashed line: RPA with GW energies; solid line: Bethe Salpeter with GW energies; circles: experiment (from [20]).

4.2. *The Time Dependent Density Functional Approach*

Various examples of calculations for the absorption spectra using the Bethe Salpeter approach can be found in the literature [1,19]. To summarize, in order to compute the excitonic spectra we need:

- A ground state self-consistent DFT calculation:
  Kohn Sham eigenstates and energies are calculated
- A GW calculation to correct the Kohn-Sham eigenvalues
- The construction and diagonalization of the effective two particle Hamiltonian.

The three steps described are of increasing difficulty. It is hence clear that the BSE approach is quite difficult and can not be applied to very complex systems.

On the other hand, DFT has shown that a theoretical description of the many-electron ground state is possible also for very large systems (hundreds of atoms). But, as already mentioned, DFT can not reproduce some quantities such as electronic excitations energies and optical spectra, being a ground state theory. A rigorous generalization of DFT to time dependent external fields (and hence able to explore the excitation spectra) was proposed by Runge, Gross and Kohn [21,22]. In their work Runge and Gross developed a theory similar to the Hohenberg-Khon-Sham one has been developed, but for time-dependent potentials. [23,24,25].

In TDDFT, as in DFT, the relevant physical quantity is the charge density n, that now is not only a function of space, but also of time. In contrast to DFT, in TDDFT, no energy minimum principle is available and the evolution of the system is described by the quantum-mechanical action, defined by:

$$A[\Psi] = \int_{t_0}^{t_1} dt < \Psi(t)|i\frac{\partial}{\partial t} - H(t)|\Psi(t) > . \tag{18}$$

For static DFT the stationary points that give the exact density of the system $n(\mathbf{r})$ are those for which the total energy obey to:

$$\frac{\delta E[n]}{\delta n(\mathbf{r})} = 0 \tag{19}$$

instead in TDDFT these points are those for which:

$$\frac{\delta A[n]}{\delta n(\mathbf{r},t)} = 0 \tag{20}$$

holds. Furthermore, as in the static DFT, it is possible to write a set of single particle equations (the Kohn-Sham equations) from which it is possible to construct the exact density of the interacting system:

$$[-\frac{1}{2}\nabla^2 + v_{eff}(\mathbf{r},t)]\phi_i(\mathbf{r},t) = i\frac{\partial}{\partial t}\phi_i(\mathbf{r},t) \tag{21}$$

where we assume the existence of a potential $v_{eff}(\mathbf{r},t)$

$$v_{eff}(\mathbf{r},t) = v_{ext}(\mathbf{r},t) + \int \frac{n(\mathbf{r},t)}{|\mathbf{r}-\mathbf{r}'|} d\mathbf{r}' + v_{xc}(\mathbf{r},t) \tag{22}$$

for an independent particle system whose orbitals $\phi_i(\mathbf{r},t)$ yield the same charge density $n(\mathbf{r},t)$ of the interacting system,

$$n(\mathbf{r},t) = \sum_{i=1} f_i |\phi_i(\mathbf{r},t)|^2. \tag{23}$$

The effective time-dependent potential v_{eff} is the sum of the external potential, the Hartree potential and the (unknown) exchange-correlation potential. In order to calculate the optical spectra, we have to compute the microscopic dielectric function (see also eq. (8):

$$\epsilon^{-1}(\mathbf{r},\mathbf{r}',\omega) = \delta(\mathbf{r}-\mathbf{r}') + \int d\mathbf{r}'' v(\mathbf{r}-\mathbf{r}'')\chi(\mathbf{r}'',\mathbf{r}',\omega) \tag{24}$$

where v is the bare coulomb interaction and χ is the reducible polarizability. In TDDFT χ is given by:

$$\chi(\mathbf{r},\mathbf{r}',\omega) = \chi_0(\mathbf{r},\mathbf{r}',\omega) + $$

$$+ \int d\mathbf{r_1} d\mathbf{r_2} \chi_0(\mathbf{r},\mathbf{r_1},\omega) \left[\frac{1}{|\mathbf{r_1}-\mathbf{r_2}|} + f_{xc}(\mathbf{r_1},\mathbf{r_2},\omega) \right] \chi(\mathbf{r_2},\mathbf{r}',\omega) \tag{25}$$

where χ_0 is the Independent-Particle or RPA Polarizability.
The exchange and correlation kernel f_{xc} is a very complex quantity which includes all many body effects and is related to the time dependent exchange-correlation potential by.

$$f_{xc}(\mathbf{r},t,\mathbf{r}',t) = \frac{\delta v_{xc}[n(\mathbf{r},t)]}{\delta n(\mathbf{r}',t')} |_{v_{ext}=0} \tag{26}$$

Since E_{xc} (and hence v_{xc}) is unknown, it is necessary to use approximations for f_{xc}.
The lowest level of approximation consists in neglecting the f_{xc} kernel and corresponds to the Random Phase Approximation (RPA) which, as we have already seen, would correspond to neglect all many-body effects.
A better although still simple approximation is nowadays largely in use, especially in the chemistry community, based on the so called Adiabatic Local Density Approximation (ALDA). It is in fact well known that one of the most used and successful scheme for calculating ground state properties of complex systems within DFT is the Local Density Approximation, that is based on the replacement of the (unknown) exact exchange and correlation energy E_{xc} of a system of interacting electrons with the functional form of the exchange and correlation energy of an homogenous electron gas of

density $n(\vec{r})$ (see eq. 4). The LDA approximation is known to work very well even in systems far from being homogenous, and is at the basis of the huge success DFT has had in the last decades. It is hence natural to try to apply the same scheme also in the case of Time-Dependent Density Functional Theory, by approximating the kernel $f_{xc}(\vec{r}, \vec{r'}, \omega)$ by the ω independent functional derivative of the LDA exchange-correlation potential:

$$f_{xc}(\vec{r}, \vec{r'}) = \delta(\vec{r} - \vec{r'}) \frac{\delta V_{xc}^{LDA}(n(\vec{r}), \vec{r})}{\delta n(\vec{r})}. \tag{27}$$

Many calculations have been performed within the ALDA scheme, and some successes have been obtained[26]. Unfortunately, the successful applications are limited to *finite* systems (atoms, molecules, small clusters), whereas no good agreement with experiments is found for *infinite* systems (bulks, surfaces, and so on). The reason for that was found in the fact that the full TDDFT kernel $v - f_{xc}$ accounts, through to the term v (bare Coulomb interaction), of the local field effects, that are very important in clusters (since clusters are highly non-homogeneous systems, being described as matter surrounded by a lot of vacuum). The term v alone is often able to correct most of the discrepancies found between RPA-LDA and experiments in clusters, and in some cases a static and local f_{xc} kernel (as the LDA one) is able to account for the remaining excitonic effects. On the other hand, in bulks and surfaces local fields effects are often not important, hence the details of the exact f_{xc} may become essential. More sophisticated kernels, obtained from a direct comparison with the many body perturbation theory are in this case needed (see [1,27,28])

As an example for ALDA calculations, we show in Fig. 5 the optical spectra of a small silicon hydrogenated cluster, the Si_5H_{12}. Different level of approximations have there been used in the calculations: the independent particle approximation (RPA) without Local Fields effects (NLF), with Local Field effects (RPA-LDA LF), and the Adiabatic LDA (ALDA). As we can see, the optical gap in RPA-LDA without Local Fields effects is about 5.6 eV, and heavily underestimates the experimental value of 6.5 eV [29]. The ALDA value 6.5 eV is instead in very good agreement with the experiment, but it is worth to note that already the RPA-LDA LF approximation gives an almost identical spectrum as the ALDA. This confirms that for finite systems the ALDA kernel is often dominated by the local field term v. In other words, the success of the ALDA is mainly due to a correct inclusion of local fields effects and not to a good approximations to f_{xc}.

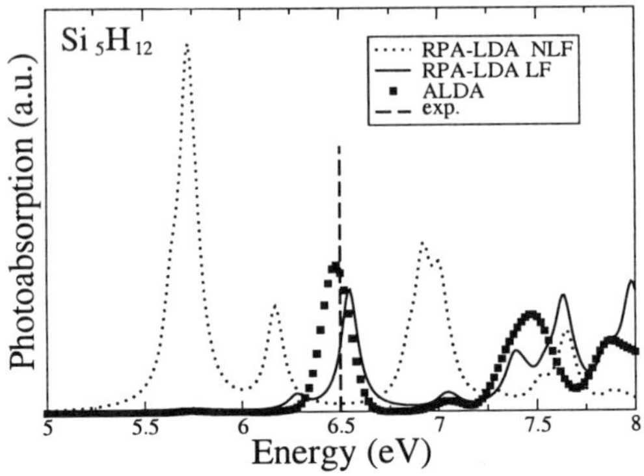

Figure 5. Comparison between the Si_5H_{12} TDLDA and LDA optical absorption spectrum

5. Conclusions

We have shown different theoretical approaches with different complexity and efforts which have to be used according to the system and to the physical aspect one wants to analyze: DFT for ground state properties, GW for band structure calculations. In order to calculate absorption spectra where excitonic effects are important, the Bethe Salpeter approach is the most refined technique, but TDDFT-ALDA represents a valid (and simpler) alternative to study isolated systems even if most of its success relies on the correct inclusion of local fields effects and not in a good approximation for the exchange and correlation TDDFT kernel.

Acknowledgments

This work has been supported by the INFM PAIS project "CELEX", and by the EU through the NANOQUANTA Network of Excellence (Contract No. NMP4-CT-2004-500198). We acknowledge CINECA CPU time granted by INFM.

References

1. for a review, see for example G. Onida, L. Reining, and A. Rubio, Rev. Mod. Phys. **74**, 601 (2002)
2. P. Hohenberg and W. Kohn, Phys. Rev. **136**, B864 (1964).
3. R.M. Dreizler and. E.K.U. Gross *Density Functional Theory* (Springer Verlag Hedelberg, 1990) R.O. Jones and O. Gunnarsson Rev. Mod. Phys. **61**, 689 (1989)
4. W. Kohn and L. J. Sham, Phys. Rev. **140**, A1113 (1965).

5. D.M. Ceperley and B.J. Alder, Phys. Rev. Lett. **45**, 566 (1980); J. P. Perdew and A. Zunger, Phys. Rev. B **23**, 5048 (1981).
6. GW fundamental gaps have been calculated within the perturbative approach, using the plasmon pole approximation. We have used 120 empty bands and 169 plane waves (411 for C and SiC).
7. Experimental gaps are taken from Landolt-Börnstein *Numerical Data and Functional Relationships in Science and Technology*, New series, Ed. O. Madelung, Springer-Verlag Berlin (1982).
8. L. Fetter and J.D. Walecka *Quantum theory of Many Body Systems* (McGraw-Hill, New York, 1981)
9. L. Hedin, Phys. Rev. **139**, A796 (1965). R. D. Mattuck, *A guide to Feynman diagrams in the many body problem*, (McGraw-Hill, New York, 1976)
10. L. Hedin and B.J. Lundquist in Solid State Physics edited by H. Ehrereich, F. Seitz and D. Turnbull (Academic press, New York, 1969), Vol **23**, p. 1
11. F. Aryasetiawan and O. Gunnarsson, Rep. Prog. Phys. **61**, 237 (1998). Aulbur W. G. Johsson and Wilkins J.W. in Solid State Physics edited by H. Ehrenreich and F. Spaepen (New York Academic, 1999)
12. R. Graupner et al., Phys. Rev. B **55**, 10841 (1997)
13. M. Marsili, O. Pulci, F. Bechstedt,R. Del Sole, Phys. Rev. **B** (accepted)
14. F.J. Himpsel, D.E. Eastman, P.Heimann, J.F. van der Veen, Phys. Rev. B **24**, 7270 (1981)
15. S. L. Adler, Phys. Rev. **126**, 413 (1962); N. Wiser, Phys. Rev. **129**, 62 (1963).
16. G. Grosso and G. Pastori Parravicini, *Solid State Physics* (Academic Press, 2000)
17. M.S. Hybertsen and S.G. Louie, Phys Rev B **34**, 5390 (1986)
18. F. Arciprete, C. Goletti, E. Placidi, P. Chiaradia, M. Fanfoni, F. Patella, C. Hogan, and A. Balzarotti, Phys. Rev. B **68**, 125328 (2003); F. Arciprete, C. Goletti, E. Placidi, C. Hogan, P. Chiaradia, M. Fanfoni, F. Patella, and A. Balzarotti, ibid., **69**, 081308 (2004)
19. G. Onida, L. Reining, R.W. Godby, R. Del Sole, and W. Andreoni, Phys. Rev. Lett. **75**, 818 (1995); M. Rohlfing and S. G. Louie, Phys. Rev. Lett. **80**, 3320 (1998); S. Albrecht, L. Reining, R. Del Sole, and G. Onida, Phys. Rev. Lett. **80**, 4510 (1998); L. X. Benedict, E. L. Shirley, and R. B. Bohn, Phys. Rev. Lett. **80**, 4514 (1998); M. Rohlfing and S. G. Louie, Phys. Rev. Lett.**82**, 1959 (1999) P.H. Hahn, W.G. Schmidt, F. Bechstedt, Phys. Rev. Lett. **88**, 016402 (2002); P. H. Hahn et al., Phys. Rev. Lett. **94**, 037404 (2005)
20. D.M. Roessler, W.C. Waljer, J. Opt.Soc. Am. **57**, 835 (1967)
21. E. Runge and E.K.U. Gross, Phys. Rev. Lett. **52**, 997 (1984)
22. E.K.U. Gross and W. Kohn, Phys. Rev. Lett. **55**, 2850 (1985)
23. E.K.U. Gross, F.J. Dobson and M. Petersilka *Density Functional Theory*, (Springer, New York, 1996)
24. M.E. Casida *Recent Developments and Applications of Modern Density Functional Theory*, edited by J.M. Seminario (Elsevier Science, Amsterdam, 1996)
25. J. Dobson, G. Vignale and M.P. Das *Electronic Density Functional Theory: Recent Progress and New Directions* (Plenum, New York, 1997)
26. see for example J.R. Chelikowsky, L. Kronik, I. Vasiliev, J. Phys.: Condens Matter **15**, R1517-R1547 (2003)

27. G. Adragna, R. Del Sole, and A. Marini Phys. Rev. B **68**, 165108 (2003)
28. F. Sottile, V. Olevano, and L. Reining Phys. Rev. Lett. **91**, 056402 (2003) The ABINIT software project (URL http://www.abinit.org)
29. Molekülspektroskopishe Untersuchungen auf dem Gebiet der Silane und der Heterocyclischen Sulfane, Forschungsbericht des Landes Nordrhein-Westfalen (West-Deutscher Verlag, Köln,1977))

THEORY OF SURFACE SECOND HARMONIC GENERATION

W. LUIS MOCHÁN AND JESÚS A. MAYTORENA*

Centro de Ciencias Físicas, Universidad Nacional Autónoma de México, Apartado Postal 48-3, 62251 Cuernavaca, Morelos, México

We develop the theory of the optical second harmonic generation from the surface and bulk of centrosymmetric systems, we discuss its advantages as an optical probe of surfaces and buried interfaces and we illustrate its fundamental aspects through simple model calculations. We apply the theory to simple isotropic harmonic surfaces, homogeneous metals, crystalline semiconductors, and nanostructured materials, and we discuss a novel technique to enhance the SHG signal.

1. Introduction

Maxwell's equations are a set of linear differential equations, which therefore seem consistent with the *superposition principle*, i.e., the linear combination of two solutions ought to be a solution. In consequence a beam of light is unaffected by the presence of others; *light is transparent.* However, these statements are not entirely correct: Maxwell equations do not form a complete set of equations within matter, as they have to be complemented by the *constitutive* equations which relate the response of the system to the field that excites it. The latter might be nonlinear and might yield a matter mediated interaction between photons. To understand these nonlinearities, consider Fig. 1a, which illustrates the fact that within materials photons are continually being absorbed, exciting electron-hole (e-h) pairs which later recombine, thus re-creating the photons; the normal propagating modes, i.e., the *polaritons*, are photons *dressed* by e-h pairs. At high intensities it is likely that an electron excited into a state $|1\rangle$ by a given photon A interacts with another photon B which further excites it into a higher energy state $|2\rangle$ before it has had time to recombine with the hole

*currently at Centro de Ciencias de la Materia Condensada, UNAM, Apdo. Postal 2681, 22800 Ensenada, Baja California, México.

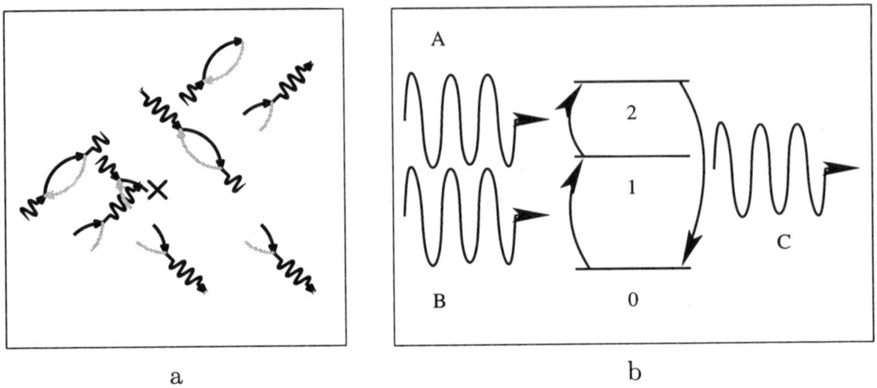

Figure 1. (a) Dressed photons (wavy arrows) propagating within a material (square). The photons excite electron- (black arrows) hole (gray arrows) pairs that recombine to produce photons. At high intensities it is likely that a photon may collide with an electron before it recombines (X). (b) The same process in terms of (virtual) transitions.

left behind in the initial state $|0\rangle$ (Fig. 1b), so that upon recombination, the emitted photon C has the combined energy of the two incoming photons, $\hbar\omega_C = \hbar\omega_A + \hbar\omega_B$, as if A and B had collided with each other producing C as an end result. This is a particular second order nonlinear process called *sum frequency generation.* The case in which A and B have the same frequency so that the outgoing photon has twice the frequency of the incoming photons is called *second harmonic generation* (SHG).[1]

SHG has become a powerful tool for the study of surfaces for the main reason that is is strongly suppressed within the bulk of a large class of systems, namely, those with a center of inversion symmetry.[2,3,4] In this paper we present simple models and theoretical aspects of SHG from centrosymmetric systems with a few applications. The paper is organized as follows. In Sec. 2 we discuss why SHG is useful as a surface and interface probe. In Sec. 3 we make a phenomenological estimate of the intensity of the expected signal. Sec. 4 illustrates a quantum-mechanical scheme for the microscopic calculation of the SH susceptibilities while Sec. 5 shows how symmetry properties may be employed to identify their relevant components. The physical interpretation of the most conspicuous ones is offered in Sec. 6. Simple analytically solvable models are presented in Sec. 7, and they are used to introduce the calculation of the SH efficiency in Sec. 8. Some refinements of the simple models are undertaken in Sec. 9: the surface local field effect and the crystalline geometry is incorporated in a polarizable-bond model and the surface localized collective modes such as the multipolar surface plasmon of metals are incorporated in a continuous hydrodynamic model. Extensions of the theory to non-flat surfaces such as those of nanoparticles embedded within a host are taken up in Sec. 10. Finally, we summarize our results in Sec. 11.

2. SHG and Surfaces

The SH radiation at frequency 2ω is produced by an oscillating polarization $\vec{P}_{2\omega}^{(2)}$ proportional in the lowest order to the square of the fundamental electric field $\vec{E}_\omega$. Thus we write $\vec{P}_{2\omega}^{(2)} = \chi^{(2)}\vec{E}_\omega\vec{E}_\omega$. The second order susceptibility $\chi^{(2)}$ is a third rank tensor, as it produces a vector when acting on two vectors. Applying an inversion, this equation becomes $-\vec{P}_{2\omega}^{(2)} = \chi_I^{(2)}(-\vec{E}_\omega)(-\vec{E}_\omega)$, where the direction of all vectors was reversed and where $\chi_I^{(2)}$ is the nonlinear susceptibility of the inverted system. However, if the system is centrosymmetric, all of its properties, and in particular, its quadratic susceptibility, are invariant under inversions. Therefore, $\chi_I^{(2)} = \chi^{(2)}$, $P_{2\omega}^{(2)} = -P_{2\omega}^{(2)} = 0$ and there is no SHG from the bulk of centrosymmetric systems. There is no finite third order tensor invariant under inversions.

The result above may be understood by recalling that the states of a centrosymmetric system may be classified according to their parity. If the transitions in Fig. 1b were driven by the dipolar interaction Hamiltonian $H_{\text{int}} = -\hat{p}\cdot\vec{E}$, then $\chi^{(2)}$ would be proportional to the probability amplitude for the chain of transitions $|0\rangle \to |1\rangle \to |2\rangle \to |0\rangle$, i.e., $\chi^{(2)} \propto \langle 0|\hat{p}|2\rangle\langle 2|\hat{p}|1\rangle\langle 1|\hat{p}|0\rangle$. As the dipolar operator $\hat{p}$ only couples even to odd states, $\chi^{(2)}$ would be different from zero only if $|0\rangle$ and $|2\rangle$ had opposite parities, as well as $|2\rangle$ and $|1\rangle$, and $|1\rangle$ and $|0\rangle$. This would require $|0\rangle$ to have the opposite parity as itself, a clear impossibility. Thus, $\chi^{(2)} = 0$.

However, surfaces are not centrosymmetric, as an inversion would interchange the inside with the nonequivalent outside. Therefore, SHG may be produced in centrosymmetric systems, but only close to their surface, within a region with a width typically of atomic dimensions, a few Å where the electrons *feel* the non-centrosymmetry of the surface ambient (Fig. 2a). Actually, there might be an additional bulk contribution to SHG but it is of a non-dipolar origin. The electric quadrupole couples to the gradient of the electric field and the quadrupolar operator doesn't change the parity. Similarly, the magnetic dipole couples to the magnetic field which is proportional to the curl of the electric field, and also leaves the parity unchanged. Thus, if one of the three transitions in Fig. 1b is multipolarly driven, it

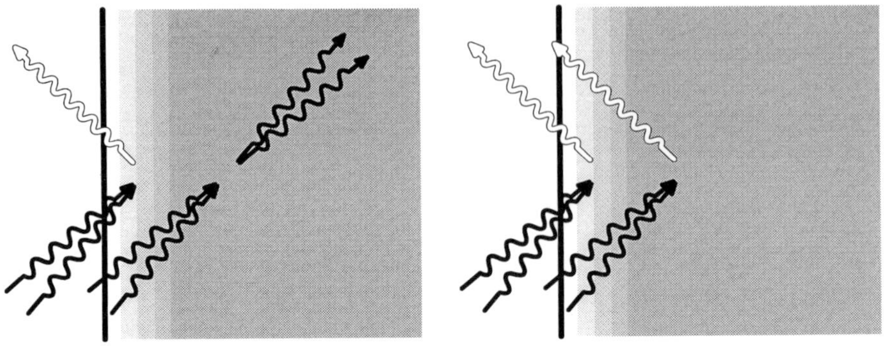

a b

Figure 2. SH photons (hollow wavy arrows) produced by the collision of fundamental photons (black wavy arrows). (a) SHG is produced close to the surface of centrosymmetric systems but not at its bulk. (b) A small multipolar contribution may also be produced at the bulk.

may induce a nonlinear response $P_i^{(2)} = \chi_{ijkl}^{(2)} E_j \partial_k E_l$ where $i \ldots l$ denote cartesian coordinates, ∂_k abbreviates the partial derivative with respect to the k-th coordinate, and we hid the frequency arguments. The fourth order rank tensor $\chi_{ijkl}^{(2)}$ is in general different from zero even within the bulk of centrosymmetric systems, as the RHS of the previous equation has three vectors (two electric fields and a gradient) and therefore changes sign upon an inversion, as does its LHS. As the gradient of the bulk field is of order E/λ, from dimensional arguments it is expected that the polarization at the bulk would be about a_B/λ times smaller than at the surface, where a_B is the Bohr radius and λ is the wavelength, but the radiating bulk region is about λ/a_B times larger, so that bulk and surface contributions to SHG are about the same order of magnitude. In contrast, the linear response is dominated by the bulk contribution and sophisticated techniques, such as reflection anisotropy spectroscopy (RAS) or differential reflectance spectroscopy (RDS) have had to be developed to eliminate the bulk signal.[5,6]

3. Intensity of Surface SHG

To estimate the strength of surface SHG we recall that a dipole $\vec{p}$ produces a near field $E \sim p/r^3$ at a distance r. As the radiation field necessarily decreases only as $1/r$ and the only other relevant distance is the wavelength λ, the far field produced at $r \gg \lambda$ by the dipole is $E \sim p/(\lambda^2 r)$. As the volume polarization and the electric field have the same units (in the Gaussian CGS system), the quadratic surface susceptibility has the units of an inverse electric field. A typical field within matter is that given by the action of the nucleus on the exterior electrons or the interaction between two electrons, so that its size is of the order of e/a_B^2. Therefore, the quadratic polarization is of order $P^{(2)} \sim E^2 a_B^2/e$. However, due to centrosymmetry, this polarization is induced only within a thin region of width $s \sim a_B$ (Fig. 3a). Different parts of this region contribute with different phases to the

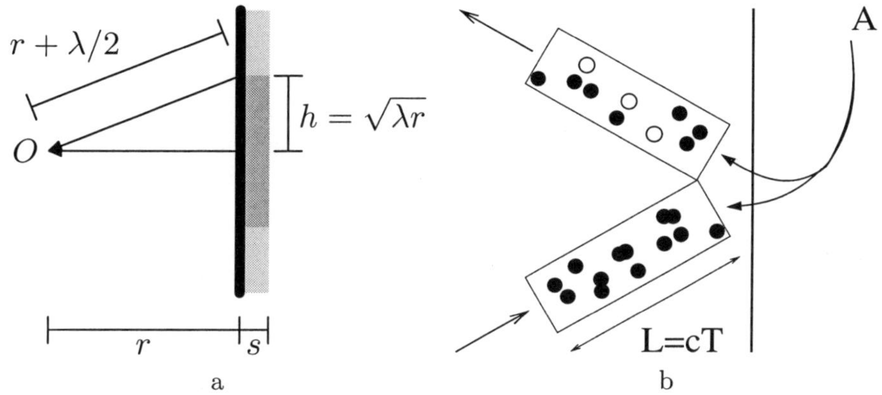

Figure 3. (a) Region (gray) close to the surface (black) where a nonlinear polarization $P^{(2)} \sim E^2 a_B^2/e$ is produced due to the surface breaking of centrosymmetry. The cylinder of radius h which contributes in phase to the field at the observation point O a distance r away is indicated in dark gray. (b) Fundamental photons (black circles) and SH photons (open circles) within a cilinder of length $L = cT$ and section A before and after being reflected by the surface. Pairs of photons that overlap arrive at similar positions on the surface at similar times and are converted to the SH.

field at the observation point, so that destructive interference eliminates all but the contributions of those regions whose distance to the observation point are between r and $\sim r + \lambda/2$. The radius of this region is $h \sim \sqrt{r\lambda}$, its volume $V \sim a_B h^2 = a_B r \lambda$ and its corresponding nonlinear dipole moment $p^{(2)} = P^{(2)} V \sim E^2 \lambda r a_B^3/e$, so that the nonlinear field at r is $E^{(2)} \sim p^{(2)}/(\lambda^2 r) \sim (a_B^3/\lambda e)E^2$. The non-linear intensity $I^{(2)} \sim c|E^{(2)}|^2$ is therefore proportional to the fourth power of the linear electric field. We define the SHG efficiency $\mathcal{R} = I^{(2)}/I^2$, where $I \sim c|E^2|$ is the intensity of the fundamental beam. Thus we finally find

$$\mathcal{R} \sim \left(\frac{a_B}{\lambda}\right)^2 \frac{a_B}{e^2} \frac{a_B}{c} a_B^2. \tag{1}$$

Notice that the efficiency grows as the square of the frequency. Assuming $\lambda \approx 10^3 a_B$ yields $\mathcal{R} \sim 10^{-28}\mathrm{cm}^2/W$. A more careful calculation (see for example our Eq. (33)) shows that our estimate above has to be modified by multiplying it by prefactors made up of fairly large powers of factors which, although they are of order 1, they bring up the non-resonant efficiency by several orders of magnitude. Simply changing the typical distance from a_B to $2a_B$ would yield an increment of two orders of magnitude. Typically, $\mathcal{R} \sim 10^{-24} - 10^{-20}\mathrm{cm}^2/W$, and resonances would increase the efficiency even more.

We can present the efficiency in an alternative way by noticing that in order for two fundamental photons to interact with each other to produce a SH photon, they would have to arrive onto the surface at nearly the same position and time. Thus we may define a photon-photon *scattering volume* in a similar fashion in which a scattering section is defined for collisions with static targets. Consider a cylindrical volume $V = AcT$ of section A and height $L = cT$ containing N fundamental photons (Fig. 3b). As it contains an energy $N\hbar\omega$ that takes a time T to cross a fixed surface, the fundamental intensity is simply $I = N\hbar\omega/(AT)$. Assume momentarily that photons may be represented by small particles (Fig. 3b) and that overlapping particles are converted to the SH upon reflection, as they arrive at the surface at close positions and times. The probability that one of these particles overlaps another within a given volume V is simply the ratio of its volume Ω to the total volume V. Thus, the number of *pairs* of photon that overlap is $N^{(2)} \sim N^2\Omega/V$. As the second harmonic intensity is $I^{(2)} = N^{(2)}2\hbar\omega/AT$, we may find the scattering *volume* of a photon, $\Omega \sim (N^{(2)}/N^2)V \sim \hbar c\omega\mathcal{R}$. Substituting Eq. (1) we obtain

$$\Omega = \frac{\hbar c}{e^2}\left(\frac{a_B}{\lambda}\right)^3 a_B^3. \tag{2}$$

Assuming again $\lambda \sim 10^3 a_B$, we obtain that the surface SH scattering volume of a photon is about 10^{-7} times the volume of an atom. Thus, surface SHG is an extremely *inefficient* process, even after accounting for the large prefactors mentioned above. However, it has the enormous advantage of being surface sensitive.

As a worked out example, consider a typical Ti:S laser which produces 200fs pulses of energy 0.3μJ focused into a spot of size 10μm with a repetition rate of 250kHz. The fundamental intensity within a pulse is $\approx 10^{13}\mathrm{W/cm}^2$, so that the SH intensity is $\approx 10^2\mathrm{W/cm}^2$. Multiplying by the pulse area and duration we obtain the energy per pulse $\approx 10^{-24}$J, which corresponds to $\approx 10^{-5}$ photons per pulse, or about 3 photons per second!

4. Quantum Mechanical Theory of the Susceptibility

To illustrate the general scheme for the calculation of the nonlinear susceptibilities[7] we start with the simple case of a single molecule. Within the *interaction picture*, its coupling with the time dependent electric field is $\hat{H}^I(t) = -\hat{p}_i(t)E_i(t)$, where $\hat{p}_i(t)$ is the i-th cartesian component of the dipolar operator, which evolves in time according to the non-perturbed Hamiltonian of the system $\hat{H}_0$, and a sum over i is implied. The density matrix $\hat{\rho}(t)$ of the system obeys a Schrödinger-like equation $i\hbar\partial\hat{\rho}(t)/\partial t = [\hat{H}^I(t), \hat{\rho}(t)]$ which may be written as a recursive equation

$$i\hbar\partial\hat{\rho}^{(n+1)}(t)/\partial t = [\hat{H}^I(t), \hat{\rho}^{(n)}(t)], \tag{3}$$

using perturbation theory, where n denotes succesive powers of the perturbing field. The zeroth order time-independent solution $\rho^{(0)}$ characterizes the equilibrium occupancy of the electronic states and is given by a thermal distribution at the equilibrium temperature. The first order solution

$$\hat{\rho}^{(1)}(t') = \frac{i}{\hbar}\int_{-\infty}^{t'} dt''[\hat{p}_k(t''), \hat{\rho}^{(0)}]E_k(t'') \tag{4}$$

may be substituted into the RHS of Eq. (3) and integrated to obtain the second order density matrix

$$\hat{\rho}^{(2)}(t) = \frac{i}{\hbar}\int_{-\infty}^{t} dt'[\hat{p}_j(t'), \hat{\rho}^{(1)}(t')]E_j(t'). \tag{5}$$

With the density matrix we may calculate the expectation value $p_i(t) = \langle\hat{p}_i(t)\rangle = \mathrm{tr}(\hat{p}_i(t)\hat{\rho}(t))$ of the dipole moment. Putting all together to second order, $p_i^{(2)}(t) = \mathrm{tr}(\hat{p}_i(t)\hat{\rho}^{(2)}(t))$, we obtain

$$p_i^{(2)}(t) = -\frac{1}{\hbar^2}\int_{-\infty}^{t} dt' \int_{-\infty}^{t'} dt''\, \mathrm{tr}\Big(\hat{p}_i(t)\,[\hat{p}_j(t'),\, [\hat{p}_k(t''), \hat{\rho}^{(0)}]]\Big)E_j(t')E_k(t''). \tag{6}$$

Notice that Eq. (6) relates the quadratic linear dipole $p_i^{(2)}$ at a give time t to the product of the electric field evaluated at two previous times t' and t''. Thus, we may rewrite it as

$$p_i^{(2)}(t) = \int dt' \int dt''\, \alpha_{ijk}(t, t', t'')E_j(t')E_k(t''), \tag{7}$$

where we have identified the second order polarizability

$$\begin{aligned}\alpha_{ijk}(t, t', t'') &= -\frac{1}{\hbar^2}\mathrm{tr}\Big(\hat{p}_i(t)\,[\hat{p}_j(t'),\, [\hat{p}_k(t''), \hat{\rho}^{(0)}]]\Big)\\ &= -\frac{1}{\hbar^2}\mathrm{tr}\Big([[\hat{p}_i(t), \hat{p}_j(t')], \hat{p}_k(t'')]\hat{\rho}^{(0)}\Big).\end{aligned} \tag{8}$$

subject to the causality requirement ($\alpha_{ijk}(t,t',t'')=0$ unless $t\geq t'\geq t''$). In the last line of Eq. (8) we employed the identity $\mathrm{tr}(\hat{A}[\hat{B},[\hat{C},\hat{D}]]) = \mathrm{tr}([[\hat{A},\hat{B}],\hat{C}]\hat{D})$, which allows us to rewrite it as

$$\alpha_{ijk}(t,t',t'') = -\frac{1}{\hbar^2}\Big\langle\big[[\hat{p}_i(t),\hat{p}_j(t')],\hat{p}_k(t'')\big]\Big\rangle_0, \tag{9}$$

where $\langle\ldots\rangle = \mathrm{tr}(\ldots\hat{\rho}^{(0)})$ denotes the equilibrium expectation value of any quantity $(\ldots)$. Eq. (9) is a particular case of Kubo's formulae, which shows that the second order dipole-dipole response is simply given by the third order equilibrium correlation function of the dipolar operator. This expression may be further symmetrized in the index-variable pairs $jt'\leftrightarrow kt''$.

Eq. (9) can now be written in the frequency domain for the particular case of an adiabatically switched monochromatic field $\vec{E}(t) = \vec{E}e^{-i\omega+\eta t}$ + C.C., where $\eta\to 0^+$ is a small positive switching rate useful to give meaning to the ill defined integrals below. To that end we insert between each operator in Eq. (9) the decomposition of the unit operator $\hat{1}=\sum_\mu|\mu\rangle\langle\mu|$ in terms of the energy eigenstates $|\mu\rangle$ of $\hat{H}_0$, and use the time evolution of the dipolar operators $\langle\mu|\hat{p}_l(t)|\nu\rangle = p_l^{\mu\nu}e^{i\omega_{\mu\nu}t}$, where $p_l^{\mu\nu}$ are the dipolar matrix elements in the Schrödinger representation or equivalently, at time $t=0$, $\omega_{\mu\nu}=\omega_\mu-\omega_\nu$, and $\hbar\omega_\mu$ is the energy of $|\mu\rangle$. The result for the SH dipole is

$$\begin{aligned} p_i^{(2)}(t) = \frac{1}{\hbar^2}p_i^{0\mu}p_j^{\mu\nu}p_k^{\nu 0}\int_{-\infty}^{t}dt'\int_{-\infty}^{t'}dt''\,e^{i\omega_{0\mu}t}e^{i(\omega_{\mu\nu}-\omega)t'}e^{i(\omega_{\nu 0}-\omega)t''} \\ \times E_jE_k - \ldots \end{aligned} \tag{10}$$

which we write as $p_i^{(2)}(t) = p_i^{(2)}e^{-2i\omega t} = \alpha_{ijk}(2\omega)E_jE_ke^{-2i\omega t}$, where we identify

$$\alpha_{ijk} = -\frac{1}{\hbar^2}\frac{p_i^{0\mu}p_j^{\mu\nu}p_k^{\nu 0}}{(\omega_{\mu 0}-2\omega)(\omega_{\nu 0}-\omega)} \pm \text{permutations}. \tag{11}$$

The results above were calculated for a single non-centrosymmetric molecule. If we have n_s such molecules per unit area at the surface of a centrosymmetric solid, the nonlinear surface polarization would be $\vec{P}^{(2)} = n_s\vec{p}^{(2)}$, yielding the surface susceptibility $\chi_{ijk} = n_s\alpha_{ijk}$. For a many-electron extended system the result above would have to be modified, but its structure would be essentially the same, that is, the susceptibility would be given by a sum over sets of three states of terms, each of which contains three matrix elements of the relevant operators that couple the system to the electric field, and with quadratic energy denominators which are resonant whenever the fundamental frequency corresponds to a transition energy or to its subharmonic.[8,1]

As discussed above, the dipolar susceptibility would vanish within the bulk of a centrosymmetric medium. Nevertheless, we may calculate its multipolar response by incorporating the spatial dependence of the field in the interaction Hamiltonian.[9,10] For a single molecule, we would write $H^I = -\hat{p}_l E_l - \hat{m}_l B_l - (1/6)\hat{Q}_{kl}\partial_k E_l$ where $\hat{m}_l$ and $\hat{Q}_{kl}$ are the components of the magnetic dipole and of the electric quadrupole operators. An odd number of the transitions $|0\rangle \to |\mu\rangle \to |\nu\rangle \to |0\rangle$ should be multipolarly driven in order to have a non-null result due to parity. As $\vec{B} \propto \nabla \times E$, both $\hat{m}_l$ and $\hat{Q}_{kl}$ couple to $\partial_k E_l$. Thus, the bulk polarization may be written as

$$P_i^{(2)} = np_i^{(2)} - (1/6)\partial_j n Q_{ij}^{(2)} = \chi_{ijkl} E_j \partial_k E_l, \tag{12}$$

where n is the bulk number density of molecules and

$$\chi_{ijkl} = -\frac{n}{6\hbar^2} \frac{p_i^{0\mu} p_j^{\mu\nu} Q_{kl}^{\nu 0}}{(\omega_{\mu 0} - 2\omega)(\omega_{\nu 0} - \omega)} \pm \dots \tag{13}$$

is the bulk nonlinear susceptibility. Notice that since the bulk response depends not only on the values of the field, but also on its spatial variation, the multipolar response is intrinsically of a *non-local* or *spatially dispersive* character.

5. Symmetry

We have developed a scheme for the quantum calculation of the SH susceptibilities of centrosymmetric systems and displayed a few terms of the resulting expressions. Their use is however hampered by their intrinsic complexity, by the large number of terms that have to be included and by the large number of components of the surface (27) and bulk (81) susceptibilities. Fortunately, not all of these components are independent and their number can be greatly diminished by employing symmetry arguments.

As a worked out example, consider an isotropic surface with a mirror symmetry. That is, consider a surface that doesn't change when rotated by an arbitrary angle around its normal, which we take along the z axis and which is invariant under a reflection in any mirror that is perpendicular to the surface. In general, any symmetry operation S that doesn't modify the system should leave its response functions invariant.[8] Thus, the surface susceptibility has to obey

$$\chi_{ijk} = \chi_{ikj} = S_{ii'} S_{jj'} S_{kk'} \chi_{i'j'k'} \tag{14}$$

for all of the independent surface symmetries S, represented by the matrices $S_{ii'}$. For our example, we consider a rotation $R_{\delta\theta}$ given up to second order in the *infinitesimally* small angle $\delta\theta$ by

$$\overleftrightarrow{R}_{\delta\theta} = \begin{pmatrix} 1-\delta\theta^2/2 & -\delta\theta & 0 \\ \delta\theta & 1-\delta\theta^2/2 & 0 \\ 0 & 0 & 1 \end{pmatrix}, \tag{15}$$

a mirror reflection in the $y-z$ plane,

$$I_x = \begin{pmatrix} -1 & 0 & 0 \\ 0 & 1 & 0 \\ 0 & 0 & 1 \end{pmatrix}, \tag{16}$$

and on the $x-z$ plane,

$$I_y = \begin{pmatrix} 1 & 0 & 0 \\ 0 & -1 & 0 \\ 0 & 0 & 1 \end{pmatrix}, \tag{17}$$

although the last one is not independent of the first two. The jk permutation symmetry in Eq. (14) is due to the fact that χ_{ijk} acts on the symmetric combination E_jE_k. As χ_{zzz} is left invariant and is not mixed with other components of χ_{ijk} by the three symmetries above, it may take an arbitrary value. Acting on χ_{zzx} with I_x we obtain $-\chi_{zzx}$. Therefore, according to Eq. (14) we must have $\chi_{zzx} = 0$. Similarly, we obtain $\chi_{zzx} = \chi_{zzy} = \chi_{zxz} = \chi_{zyz} = \chi_{xzz} = \chi_{yzz} = 0$. Acting with either I_x or I_y on χ_{zxy} we obtain $-\chi_{zxy}$. Therefore, $\chi_{xyz} = 0$. Similarly, $\chi_{yzx} = \chi_{zxy} = \chi_{xzy} = \chi_{zyx} = \chi_{yxz} = 0$. Applying $R_{\delta\theta}$ to χ_{ijk} we obtain $\chi_{zxx} = (1-\delta\theta^2)\chi_{zxx} + \delta\theta^2\chi_{zyy}$, which is only possible if $\chi_{zxx} = \chi_{zyy}$, which otherwise can take any value. Similarly, $\chi_{xzx} = (1-\delta\theta^2)\chi_{xzx} + \delta\theta^2\chi_{yzy}$, from which we infer that $\chi_{xzx} = \chi_{yzy}$ is another independent contribution. There are no more constrictions due to the above symmetries, so that there are only three independent components to the surface susceptibility of isotropic non-chiral surfaces, namely, $\chi_{\perp\perp\perp}$, $\chi_{\parallel\perp\parallel}$, and $\chi_{\perp\parallel\parallel}$.

The independent non-null components of the surface susceptibility may be obtained more systematically by employing group theoretical methods.[11] For the different crystalline surfaces they are shown[4] in table 1. The surface crystaline structure is therefore reflected in the dependence of the surface polarization on the relative orientation of the polarizing field and the crystaline axes yielding an anisotropy.[12,13] As an example, in Fig. 4 we show experimental results[12] for the p polarized SHG from a Si(111) surface illuminated with s polarized light. The threefold rotation symmetry of the

Table 1. Independent non-null components[4] of the surface susceptibility χ_{ijk} for different symmetry groups, assuming that the surface normal points along z.

Symmetry	Non-null components of χ_{ijk}
1	xxx, xxy, xyy, yxx, yxy, yyy, xxz, xyz, yxz, yyz, zxx, zxy, zyy, xzz, yzz, zxz, zyz, zzz
1m ($\perp y$)	xxx, xyy, xzz, xzx, yzy, yxy, zxx, zyy, zxz, zzz
2	xzx, xyz, yxz, yzy, zxx, zyy, zxy, zzz
2mm	xzx, yzy, zxx, zyy, zzz
3	$xxx = -xyy = -yyx$, $yyy = -yxx = -xyx$, $yzy = xzx$, $zxx = zyy$
3m ($\perp y$)	$xxx = -xyy = -yxy$, $xzx = yzy$, $zxx = zyy$, zzz
4, 6, ∞	$xxz = yyz$, $zxx = zyy$, $xyz = -yxz$, zzz
4mm, 6mm, ∞m	$xxz = yyz$, $zxx = zyy$, zzz

surface is inherited by the SH intensity as a function of the angle between the plane of incidence and the crystal axes.

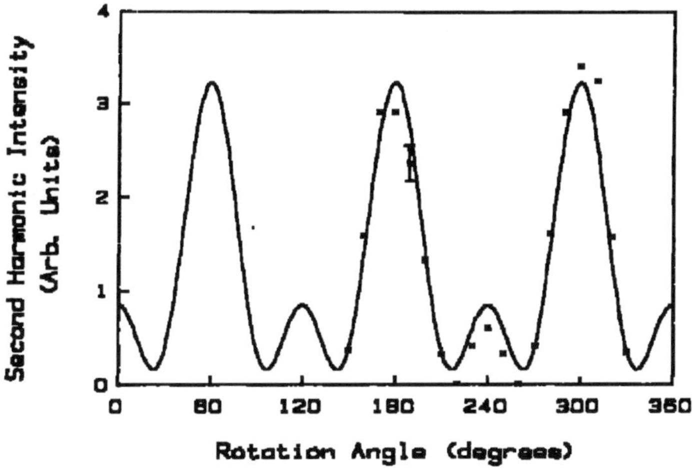

Figure 4. P polarized SH intensity produced by a Si(111) surface illuminated with s polarized light as a function of the orientation of the incidence plane.[12]

6. Physical Origin of the Nonlinear Surface Response

We may understand the physical origin of some of these components by studying the case of an isotropic metalic surface. Consider a field applied normally towards the surface of a conductor (Fig. 5a). The field is screened by a surface charge σ which, contrary to elementary textbooks assumptions,

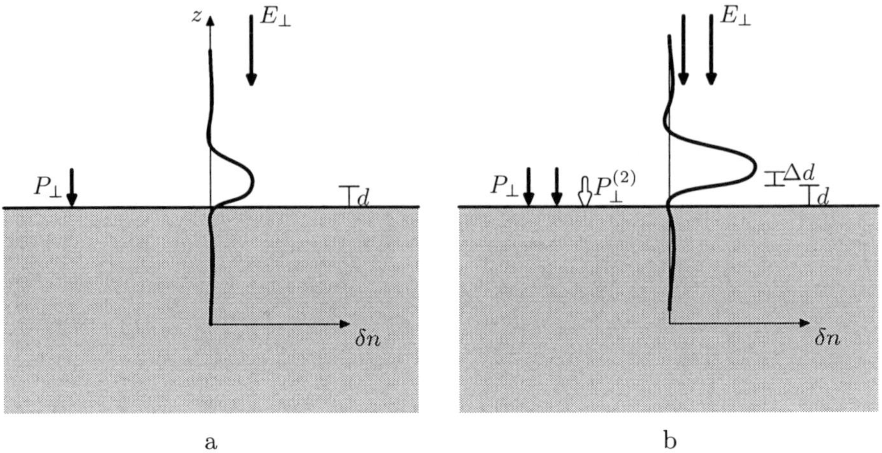

Figure 5. A normal electric field $E_\perp$ draws electrons from a conductor (gray) which accumulate at the surface with a distribution $\delta n(z)$ whose centroid is a distance d away from the nominal surface, yielding a linear surface polarization $P_\perp = \sigma d$. (b) A stronger field produces a larger surface charge and deforms its distribution, displacing its centroid by a distance Δd and producing a nonlinear perpendicular polarization $P_\perp^{(2)}$.

is not localized in an infinitesimally thin sheet, but is rather distributed in a thin region close to the surface according to a distribution $\delta n(z)$ whose centroid lies typically a small distance d outside of the nominal metal surface.[14] Thus, there is a surface polarization $P_\perp = \sigma d$ which plays some role in the linear optical response of surfaces, where $\sigma = -e \int dz\, \delta n(z)$. If the field is larger (Fig. 5b), not only is the screening charge σ increased proportionately, but the charge distribution $\delta n(z)$ is deformed and the centroid moves further away from the surface by some extra distance Δd, yielding an extra surface polarization $P_\perp^{(2)} = \sigma \Delta d$. Notice that when the fundamental field reverses direction after half a period the sign of the screening charge changes sign, but as the electrons are now strongly pushed onto the surface instead of being strongly pulled, the extra displacement Δd also changes sign so that the extra polarization regains its initial value. Thus $P_\perp^{(2)}$ oscillates with frequency 2ω around a nonzero mean value. The susceptibility that describes this process, $\chi_{\perp\perp\perp}$ is proportional to the amplitude of Δd and is usually characterized by a dimensionless parameter[15]

$$a \equiv -64\pi^2 n_B e \left(\frac{\epsilon_1^B}{\epsilon_1^B - 1} \right)^2 \chi^s_{\perp\perp\perp}, \tag{18}$$

where n_B is the bulk electronic density and ϵ_1^B the bulk linear dielectric function at the fundamental frequency ω.

In a similar way, a parallel field $E_\parallel$ may set the screening charge σ in motion along the surface a distance ℓ, yielding a parallel polarization $P_\parallel^{(2)} = \sigma\ell$. When $E_\parallel$ reverses its direction, ℓ changes sign but σ is also reversed, so that $P_\parallel^{(2)}$ takes its original value. Therefore, mixing $E_\parallel$ with $E_\perp$ yields a parallel polarization that oscillates in the second harmonic. This process is described by $\chi_{\parallel\perp\parallel}$ which is commonly characterized by the dimensionless parameter[15]

$$b \equiv -64\pi^2 n_B e \frac{\epsilon_1^B}{(\epsilon_1^B - 1)^2} \chi_{\parallel\perp\parallel}^s. \tag{19}$$

The remaining component $\chi_{\perp\parallel\parallel}$, characterized analogously by a dimensionless parameter f,[16] doesn't have such a simple interpretation, and turns out to be null for simple models that are translationally invariant in a microscopic scale, but not for crystals nor for rough surfaces.[17,18]

7. A Simple Model

In these section we present one of the most simple models for the calculation of surface SH spectra, namely, the *dipolium* model, in which a solid is assumed to be made up of a continuous distribution of harmonic polarizable entities.[19] Later we will show some refinements and generalizations of the model.

Consider a classical system made up of a charge $-e$ bound by a linear oscillator to a nucleus, which we situate at the origin, and forced by an oscilating electromagnetic field. This seems to be the most linear problem we might conceive. However, if the electromagnetic field is non-homogeneous, it is necessary to evaluate the field at the position $\vec{r}$ of the electron and not at the origin. Thus, the equation of motion for the electron is

$$m\ddot{\vec{r}} = -e\vec{E}(0,t) - m\omega_0^2\vec{r} - \frac{m}{\tau}\dot{\vec{r}} - e\vec{r}\cdot\nabla\vec{E}(0,t) - \frac{e}{c}\dot{\vec{r}} \times \vec{B}(\vec{0},t), \tag{20}$$

where the succesive terms on the RHS are respectively the driving term if the electron were at the origin, the oscillator restitutive force, a phenomenological damping term, the lowest order correction to the electric force due to the motion of the electron and the magnetic force due to the finite speed of the electron. Notice that the first three terms are the usual ones for a forced, damped, harmonic oscillator. The fourth term is linear in the displacement, so it looks as a modification to the restoring force. However, the corrections to the spring constant depends on time as $\vec{E}$ is oscillatory. Therefore, our linear oscillator has turned into a parametric oscillator, a well known nonlinear system. The last term is similar to the fourth, although linear in the velocity instead of the displacement from equilibrium. As the magnetic field is proportional to the curl of the electric field, both the fourth and fifth terms are present only for non-uniform fields.

Eq. (20) may be solved iteratively by proposing a perturbative series in powers of the driving field. The lowest order solution leads to the well known relation

$$\vec{p}^{(1)} = \alpha(\omega)\vec{E} \tag{21}$$

between the linearly induced dipole moment at the fundamental frequency and the driving field, where the polarizability $\alpha(\omega)$ is a simple Lorentzian centered at the resonance frequency ω_0,

$$\alpha(\omega) = \frac{e^2/m}{\omega_0^2 - \omega^2 - i\omega/\tau}. \tag{22}$$

Substituting the linear displacement $\vec{r}^{(1)} = -\vec{p}^{(1)}/e$ into the RHS of Eq. (20) and identifying the second order terms, we may obtain the second order displacement and polarization

$$\vec{p}^{(2)} = -\frac{1}{2e}\alpha(\omega)\alpha(2\omega)[\nabla E^2 - 4\vec{E} \times (\nabla \times \vec{E})]. \tag{23}$$

Notice that within this model, the SH response is given by products of the linear response evaluated at the fundamental and at the SH frequency. Thus, as in Eq. (11), we expect resonant behavior at the normal frequency of the system and at its subharmonic. We can also calculate the nonlinear quadrupole $\overleftrightarrow{Q}^{(2)} = -3e\vec{r}\vec{r}$ (prior to the removal of its trace), obtaining

$$\overleftrightarrow{Q}^{(2)} = -\frac{3}{e}\alpha^2(\omega)\vec{E}\vec{E}, \tag{24}$$

so that the quadrupolar response is also given in terms of the linear polarizability.

The macroscopic nonlinear polarization of a continuous solid,[20]

$$\vec{P}^{(2)} = n\vec{p}^{(2)} - \frac{1}{6}\nabla \cdot n\overleftrightarrow{Q}^{(2)}, \tag{25}$$

contains the usual dipole moment density and a correction due to the spatial variation of the quadrupole density. As $\vec{p}^{(2)}$ is proportional to $\vec{E}\nabla\vec{E}$ while $\overleftrightarrow{Q}^{(2)}$ is proportional to $\vec{E}\vec{E}$, both terms in Eq. (25) are of the same order of magnitude and proportional to $\vec{E}\nabla\vec{E}$, as opposed to the linear case where it is seldom necessary to account for the quadrupolar correction.

Recalling that the component of the displacement $D_\perp$ normal to an interface is a continuous field even when the dielectric function has abrupt discontinuities, we realize that the corresponding electric field $E_\perp$ has abrupt variations close to the surface. Therefore, the polarization $\vec{P}^{(2)}$ may take a very large value within the selvedge region across which the response of the solid changes rapidly before settling on its bulk value. Thus, from Eq. (25) we expect a very large polarization close to the surface and a relatively small value within the bulk. By linearly screening $\vec{P}^{(2)}$ and integrating it across the selvedge we may obtain a surface polarization $\vec{P}_s^{(2)}$ from which we may identify the surface susceptibility χ^s_{ijk}.

As an example, we derive $\chi^s_{\perp\perp\perp}$. First, we substitute Eqs. (23) and (24) into (25) to write the SH polarization as

$$\begin{aligned} P_z^{(2)}(z) = n(z)\alpha_2 E_z^{(2)}(z) - \frac{n(z)}{2e}\alpha_1\alpha_2\frac{\partial}{\partial z}E_z^2(z) \\ +\frac{1}{2e}\alpha_1^2\frac{\partial}{\partial z}\left(n(z)E_z^2(z)\right), \end{aligned} \tag{26}$$

where we added a contribution due to the linear response $n\alpha_2$ to the second harmonic field $\vec{E}_z^{(2)}$ and we chose the surface normal along z. Here we employed the subindices $\eta = 1, 2$ to distinguish quantities α_η and ϵ_η at the fundamental and at the SH frequencies.

Due to the strong polarization at the surface, we approximate $E_z^{(2)}$ by the depolarization field $E_z^{(2)} \approx -4\pi P_z^{(2)}$. Furthermore, $E_z(z) = D_z/\epsilon_1(z)$ where $\epsilon_\eta(z) = 1 + 4\pi n(z)\alpha_\eta$ is the position-dependent dielectric function, and D_z can be taken as position independent across the thin selvedge. Thus, Eq. (26) yields

$$P_z^{(2)}(z) = \frac{1}{2e\epsilon_2(z)}\left[-\alpha_1\alpha_2 n(z)\frac{\partial}{\partial z}\frac{1}{\epsilon_1^2(z)} + \alpha_1^2\frac{\partial}{\partial z}\frac{n(z)}{\epsilon_1^2(z)}\right]D_z^2. \tag{27}$$

As the RHS depends on z through $n(z)$, Eq. (27) may be integrated immediately to obtain the surface polarization $P_{sz}^{(2)} = \chi_{zzz}D_z^2$ and identify the parameter (18)

$$\begin{aligned} a(\omega) = 2\left([\epsilon_2^B - \epsilon_1^B][2\epsilon_1^B - \epsilon_2^B - \epsilon_1^B\epsilon_2^B]\right. \\ \left.+[\epsilon_1^B]^2[1-\epsilon_2^B]\log[\epsilon_1^B/\epsilon_2^B]\right)/[\epsilon_2^B - \epsilon_1^B]^2, \end{aligned} \tag{28}$$

which is an analytic function of the bulk dielectric function ϵ_1^B and ϵ_2^B at the fundamental and SH frequencies and is independent of the details of the number density profile $n(z)$ of polarizable entities. This expression may be employed to estimate the SH response of arbitrary systems in terms of their linear response. It does take into account the strong field variations at the surface, although it ignores surface specific features such as surface states, surface modified bulk transitions and the surface local field effect. Similarly, the dipolium model yields the value $b = -1$ (Eq. (19)) and $f = 0$ ($\chi_{\perp\parallel\parallel} = 0$). Evaluating Eq. (25) within the bulk yields the polarization

$$P^{(2)} \equiv \frac{1}{32\pi^2 ne}(\epsilon_1^B - 1)(\epsilon_2^B - 1)d(\omega)\nabla E^2 \tag{29}$$

where $d = 1$.

From the response functions above, a is the most interesting. Fig. 6 shows that for the harmonic dipolium, its imaginary part a'' is different

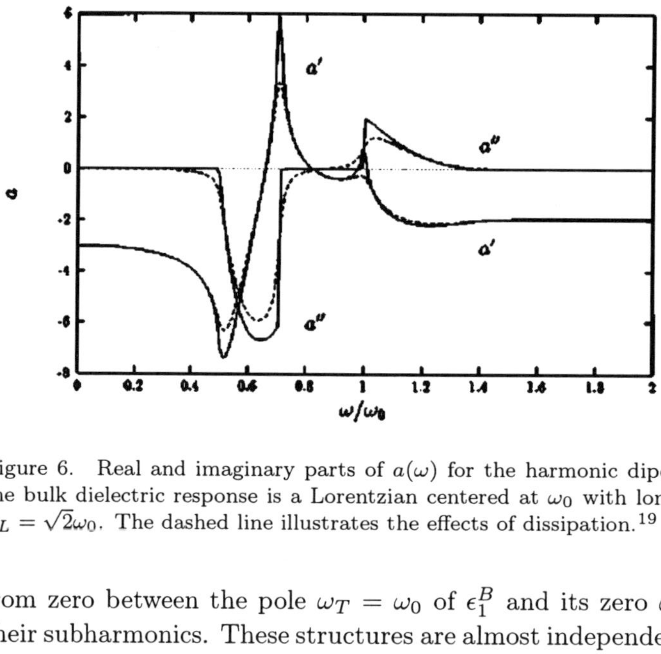

Figure 6. Real and imaginary parts of $a(\omega)$ for the harmonic dipolium model, where the bulk dielectric response is a Lorentzian centered at ω_0 with longitudinal frequency $\omega_L = \sqrt{2}\omega_0$. The dashed line illustrates the effects of dissipation.[19]

from zero between the pole $\omega_T = \omega_0$ of ϵ_1^B and its zero ω_L, and between their subharmonics. These structures are almost independent of dissipation, which merely rounds off the abrupt bends at $\omega_T/2$, $\omega_L/2$ and ω_T. The real part a' is related to a'' through causality relations and has corresponding oscillations. At high frequencies it attains the asymptotic value $a \to -2$.

8. SHG Radiation

In order to compute the efficiency[21,22,2,19] of SHG radiation, the first step that has to be taken is the solution of the linear problem, yielding the total field at the surface and at the bulk in terms of the incident field (Fig. 7a). For incoming p polarization, this is characterized by the linear transmission amplitude at the fundamental frequency t_1^p. The linear

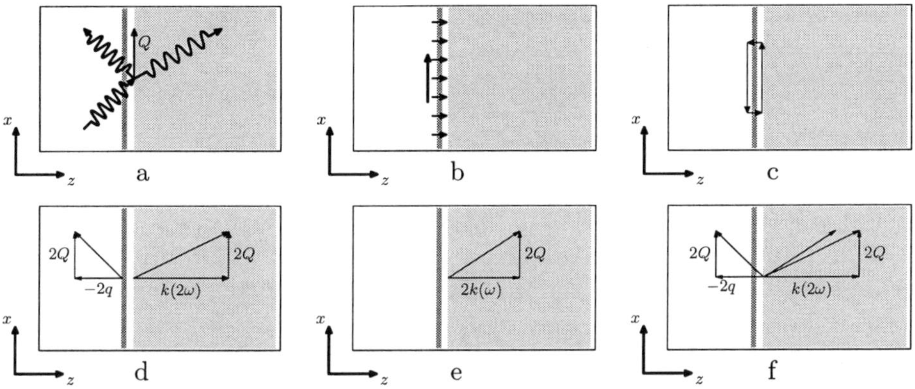

Figure 7. (a)A p polarized wave of frequency ω with wavevector $\vec{Q}$ along the surface is partially reflected and transmitted linearly with amplitude t_1^p by a solid (light gray). (b)The total linear field at the surface produces a surface nonlinear polarization (arrows) at the selvedge (dark gray) with both normal and parallel components. (c) Circuit to obtain boundary conditions across selvedge. (d)Wave vectors of surface induced SH fields within the media and vacuum. (e)Wave vectors of the SH bulk polarization source $P^{(2)}$. (f)Wave vectors of the inhomogeneous and homogeneous SH in the bulk and the bulk induced reflected SH wave.

field at the surface $E_i(0^-) \propto t_1^p$, produces a surface nonlinear polarization $P_{si}^{(2)} = \chi_{ijk}^s E_j(0^-)E_k(0^-)$ (Fig. 7b). The electromagnetic field changes from one side to the other of the selvedege due to the surface polarization and currents. Integrating Maxwell equations along a circuit such as that in Fig. 7c, we obtain the boundary conditions

$$H_y^{(2)}(0^-) = H_y^{(2)}(0^+) + 8\pi i\omega/c\, P_{sx}^{(2)} \tag{30}$$

$$E_x^{(2)}(0^-) = E_x^{(2)}(0^+) + 8\pi i Q P_{sz}^{(2)}. \tag{31}$$

As $H_y(0^+)$ and $E_x(0^+) = Z_2^p H_y(0^+)$ are related through the SH surface impedance of the medium, $Z_2^p = (k_2 c)/(2\omega\epsilon_2^B)$, and $H_y(0^-)$ and $E_x(0^-) = -Z_p^V H_y(0^+)$ are related through the vacuum surface impedance $Z_p^V = qc/\omega$, we may solve Eqs. (30) and (31) to obtain $H_y^{(2)}(0^-)$ and the surface contribution to the radiated field (Fig. 7d). Here, we introduced the normal component of the SH wave within the medium, k_2, and in vacuum, $2q$ (Fig. 7d).

On the other hand, the polarization $\vec{P}^{(2)}$ induced within the bulk may be obtained directly from the linear field using Eq. (29), which yields a wave that propagates with wavevector $2\vec{Q}$ along the surface and $2k_1$ in the normal direction (Fig. 7e), where k_1 is the normal wave vector for the fundamental field. This polarization is a source for the SH field, $\vec{E}^{(2)}$, which obeys the inhomogeneous wave equation

$$\nabla^2 \vec{E}^{(2)} + \left(\frac{2\omega}{c}\right)^2 \epsilon_2^B \vec{E}^{(2)} = -4\pi \left(\frac{2\omega}{c}\right)^2 \vec{P}^{(2)}. \tag{32}$$

Its solution has an inhomogeneous contribution $4\pi P^{(2)}/[(2k_1)^2 - k_2^2]$ and an homogeneous contribution, propagating with wave vectors $2k_1$ and k_2 respectively along the normal direction (Fig. 7f). Their electric and magnetic fields must be matched at the surface to those of the bulk-induced reflected SH field through the usual boundary conditions, from which the amplitude of the later may be calculated (Fig. 7f). Finally, the total SH field is the sum of the surface and bulk contributions.

Following the steps above for the case of an isotropic surface we obtain the SH efficiency

$$R_{pP} \equiv \frac{I^{(2)}}{I^2} = \frac{2\pi^3\omega^2}{(n_B e)^2 c^3} |r_{pP}|^2, \tag{33}$$

for the case of a p polarized incoming wave and a p polarized outgoing wave, where I and $I^{(2)}$ denote the fundamental and SH intensities, and

$$r_{pP} = \frac{Q}{q}\left(\frac{\epsilon_1^B - 1}{4\pi}\right)^2 \frac{t_2^p (t_1^p)^2}{\epsilon_2^B \epsilon_1^B} \left(\frac{\epsilon_2^B}{\epsilon_1^B}\left(\frac{Qc}{\omega}\right)^2 a \right.$$
$$\left. - \frac{k_1 k_2 c^2}{\omega^2 \epsilon_1^B} b + 2\frac{\epsilon_2^B - 1}{\epsilon_1^B - 1} d\right). \tag{34}$$

Similar formulae may be obtained for different in-out polarization combinations and for crystalline non-isotropic surfaces.[12]

As an example, in Fig. 8 we show the SH efficiency calculated with Eqs. (33) and (34) and within the dipolium model for a Si surface. Structure is clearly visible at the singular points E_1 and E_2 of the Si joint density of states and at their subharmonics.

9. Refinements

The dipolium theory above doesn't account for any specific surface phenomena, and it only incorporates the abrupt variation of the electric field

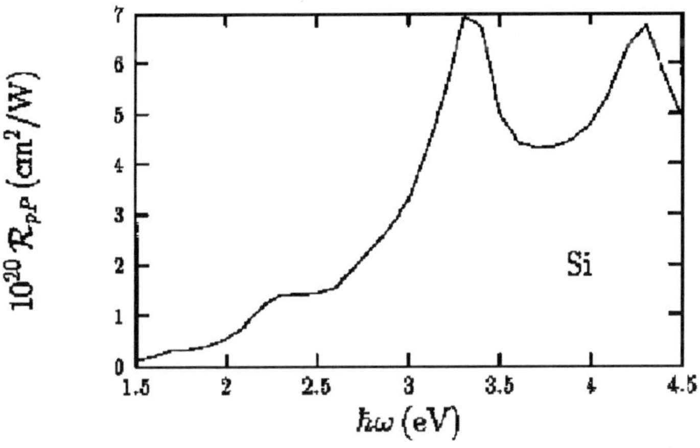

Figure 8. SH efficiency from a Si surface calculated within the continuous dipolium model as a function of the fundamental frequency.[19]

at interfaces. However, it is a convenient starting place upon which more realistic models may be built, as illustrated below.

9.1. *Surface Local Field Effect*

In contrast to the dipolium model above, the polarizable entities within a crystal would not be distributed continuously but in a lattice, and would not be polarized by the macroscopic electric field, but by the *local field*, that is, the sum of the external field and the field produced by its neighbor entities. As the latter may be at quite close distances, the field they produce is expected to have a large gradient, with a lengthscale of the order of an inverse interatomic distance. However, the gradient of the field produced by neighbors on one side is mostly cancelled out by the gradient of the field produced by equivalent neighbors on the other side within the centrosymmetric ambient of the bulk (Fig. 9). At the surface, there is no such cancellations as there are neighbors only on one side. Thus, as in the continuous dipolium model, Eqs. (23), (24), and (25) yield a large surface polarization and a small bulk polarization.

The ideas above have been applied to Si *crystals*, asuming that the Si-Si bonds are the appropriate polarizable entities, with a polarizability $\alpha_{\parallel}$ along the bond directions and a polarizability $\alpha_{\perp}$ in the normal direction.[23] From the tetrahedral geometry and fixing $\alpha_{\perp}$, $\alpha_{\parallel}$ was fitted to the dielectric response and both the linear reflection anisotropy spectra (RAS) and

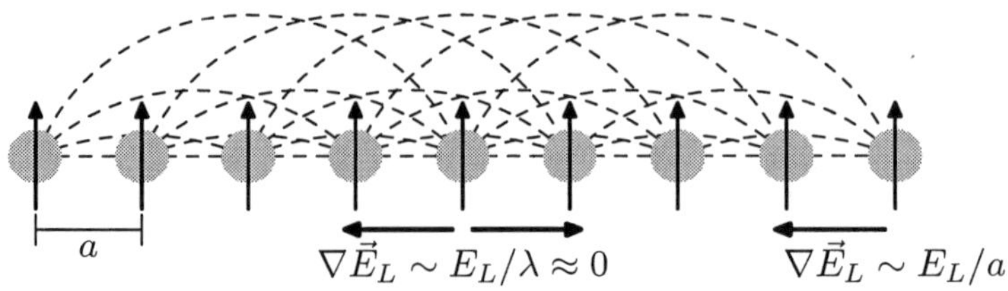

Figure 9. Lattice of polarizable entities (gray circles) whose dipole moments (narrow arrows) interact with one another (dashed lines). The gradient $\nabla \vec{E}_L$ (wide arrows) of the contributions to the local field $\vec{E}_L$ acting on an entity within the bulk cancel out, leaving a small residual gradient $\sim E_L/\lambda$ due to the wave propagation. However, the gradient of the local field at the surface $\sim E_L/a$ is not cancelled, as the ambient is not centrosymmetric.

the SHG corresponding to several crystalline surfaces was calculated for different frequencies.[23] For example, in Fig. 10 we show the anisotropy for the (111) $s \to p$ SHG. Notice the threefold symmetry, the alternating peaks due to the interference between the different contributions to SHG (table 1), and the dependence of the relative heights on ω, as the different contributions to the susceptibility have different ω behavior.

9.2. *Multipolar Plasmons at Metal Surfaces*

The continuous dipolium model may be applied to metals, simply by eliminating the harmonic restoring force $\omega_0 \to 0$. However, in a metal there are additional forces originated in the pressure of the electron gas. As electrons are fermions, they are subject to a statistical repulsion due to Pauli's exclusion principle. The simplest way to account for these is within a hydrodynamic model, which we illustrate by calculating $\chi^s_{\perp\perp\perp}$.[24] The model is defined by the continuity equation

$$\frac{\partial}{\partial t} n + \frac{\partial}{\partial z} n u_z = 0, \tag{35}$$

and Euler's equation for the momentum conservation,

$$mn \frac{\partial}{\partial t} u_z + mn \frac{u_z}{\tau} + mn u_z \frac{\partial}{\partial z} u_z = -neE_z - \frac{\partial}{\partial z} p(n), \tag{36}$$

where m is the electronic mass, u_z is the mean velociy along the surface normal of the electron gas of number density n, and $p(n) = (3/5)\gamma n^{5/3}$ is the equilibrium pressure of a degenerate fermion gas, with $\gamma = (3\pi^2)^{2/3}\hbar^2/(3m)$. The consecutive terms of Eq. (36) correspond to the inertial force, friction with the positive background, convective momentum flow,

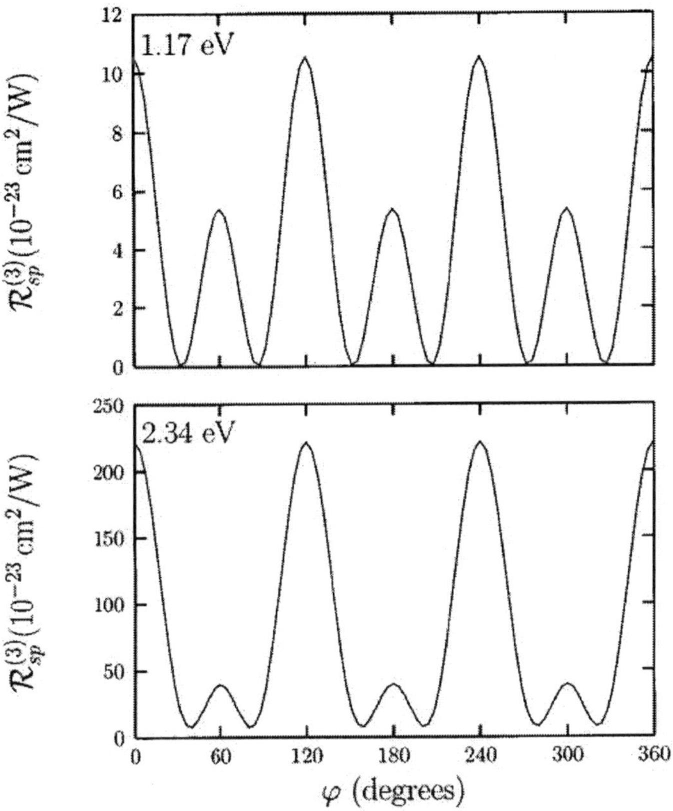

Figure 10. Dependence of the $s \to p$ SH intensity of a Si(111) surface on the azimuthal angle between the plane of incidence and the crystalline axis, calculated with the polarizable bond model at $\hbar\omega = 1.17$eV (top) and 2.34 (bottom).[23]

electric force and hydrodynamic force. Notice that the convective derivative is nonlinear in the velocity field. If we disregard the pressure term, this nonlinearity yields exactly the same results as those of the dipolium model in the limit $\omega_0 \to 0$. The pressure is also a nonlinear function of the density. Expanding all quantities in the above equations in powers of a perturbing oscillating electric field and identifying terms of the same order, we may solve the problem iteratively. The zeroth order solution yields the field $\vec{E}^{(0)}$ required to confine the electrons within the metal with a given equilibrium density profile $n_0(z)$. The profile $n_0(z)$ may be obtained from microscopical, self-consistent, quantum mechanical calculations.[25] For the hydrodynamic model, simple density profiles, linear and quadratic, were proposed. The first order equations may then be solved, yielding the surface linear polarization, which turns out to display a peak which may be identified with the *multipolar surface plasmon.*[14] This is an elusive surface-localized collective mode with a multipolar character which has been the subject of a longstanding controversy which was solved when the relatively small resonance was finally observed in electronic energy loss (EELS) experiments in alkali-metal films[26,27] and angle- and energy-resolved photoyield spectroscopy.[28] Adjusting the width of the surface density profile so that the frequency ω_m of the multipolar plasmon frequency appears close to the theoretically predicted[14] and experimentally verified frequency $\omega_m \approx 0.8\omega_p$, where ω_p is the bulk plasma frequency, all of the parameters of the theory become fixed.

The second order equations may then be solved,[24] and the SH efficiency calculated in essentially the same way as with the dipolium model above. The results displayed in Fig. 11 show very large peaks at the multipolar

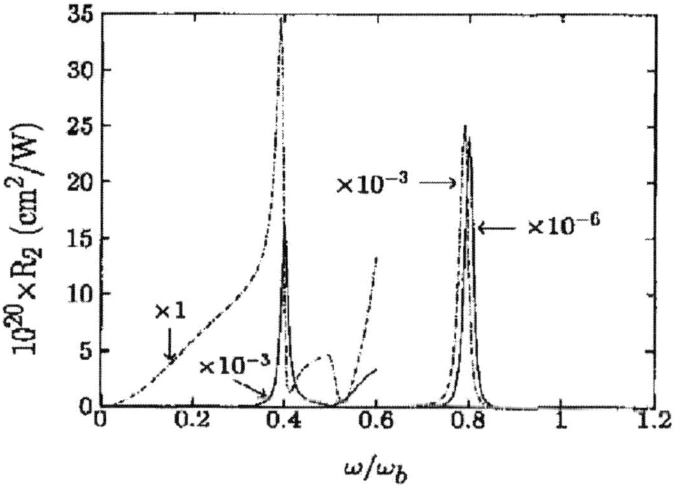

Figure 11. SHG $p \to p$ efficiency of a K surface, calculated for a linear (dashed) and a quadratic (solid) surface density profiles. The lifetime was chosen as $\tau = 30/\omega_p$ and the angle of incidence is $\theta = 60°$. In the figure R_{pP} is labeled as R_2 and ω_p is labeled as ω_b.[24]

surface plasmon frequency ω_m and at its subharmonic $\omega_m/2$, with enhancement factors between one and six orders of magnitude above the baseline, depending on the choice of profile. This result showed the feasibility of a purely optical observation of multipolar surface plasmons through SHG spectroscopy. We remark that the effect of the multipolar plasmon on surface linear optical spectra is negligible and the expected resonance in EELS is so small that very careful experiments were required to observe it. It is only very recently that the required SHG experiments were performed,[29] confirming the theoretical prediction of multipolar surface plasmon peaks at ω_m and $\omega_m/2$.

10. Nanoparticles

One of the advantages of surface- and interface-sensitive optical spectroscopies is their ability to probe buried interfaces, which are mostly inaccessible to other surface probes.[5,30] For example, SHG was recently employed to observe the Si-SiO_2 interface of Si nanocrystals embedded within a glass matrix.[31] These composite media are important technologically, as they are employed in *flash* memory devices. The nanoparticles, implanted in the insulating layer of a field effect transistor may be charged or discharged through tunneling from and to the channel when appropriate voltages are applied to the gate, and the state of the memory (charged or uncharged) may later be interrogated. The quality of the device depends on the quality of the interfaces, which have now been observed optically[31]; it has been shown that the SH signal in a transmission geometry is much stronger at the region implanted with nanoparticles than at the glass rim, that the signal is dependent on the interface condition and that the transmitted SH light is angularly distributed within a small cone and peaked close to the forward direction.

The SHG from a single nanoparticle may be understood by assuming it has a spherical shape and that the surface is *locally flat*, so that its nonlinear surface polarization may be obtained using the surface susceptibility of flat surfaces[32] (Fig. 12). For example, at position q_1 of Fig. 12, the field is normal to the surface, so that its nonlinear coupling to itself through $\chi^s_{\perp\perp\perp} \propto a$ produces a SH polarization normal to the surface. At q_3 the electric field is parallel to the surface, so $\chi^s_{\perp\parallel\parallel} \propto f$ produces a SH polarization which is also normal to the surface. On the other hand, at q_4 the electric field has a component parallel and a component perpendicular to the surface, so that $\chi^s_{\parallel\perp\parallel} \propto b$ produces a polarization parallel to the surface. However, although the centrosymmetry is locally lost at the surface of the sphere, it is globally recovered, so that when the polarization is integrated all around the sphere, the surface polarization at q_1 would be cancelled by

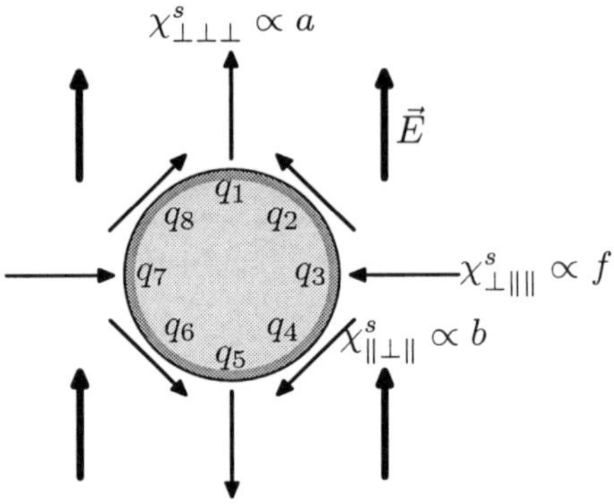

Figure 12. Applied field (wide arrows) acting on a small sphere (light gray) of a centrosymmetrical material. A SH surface polarization (thin arrows) is induced locally at the selvedge (dark gray) according to the flat-surface susceptibility χ^s_{ijk}. Some positions ($q_1 \ldots q_8$) along the surface are indicated.

the polarization at q_5, the polarization at q_7 would cancel that at q_3 and the polarization at q_2, q_4, q_6, and q_8 would add up to zero. If however, the polarizing field is inhomogeneous, the cancellation would not be exact and a residual dipole moment would appear.[33,34,35] Expanding the field in a first order Taylor series and integrating the surface and bulk polarization it induces, accounting for the dipolar and quadrupolar linear screening of the fundamental and SH fields, the total dipole moment

$$\vec{p}^{(2)} = \gamma^e \vec{E} \cdot \nabla \vec{E} + \gamma^m \vec{E} \times (\nabla \times \vec{E}) \tag{37}$$

of a single sphere may be obtained. Similarly, the quadrupole moment is

$$\overleftrightarrow{Q}^{(2)} = \gamma^q \vec{E}\vec{E}. \tag{38}$$

The phenomenological response functions γ^e, γ^m, and γ^q are given functions of the parameters a, b, f, and d discussed previously.[33,34]

Depending on the lengthscale of variation of the fundamental electric field, the SH field may be of a dipolar or a quadrupolar character, and both contributions may be comparable and interfere among themselves leading to non-trivial asymmetric radiation patterns. The wavelength of light is large, so the dipolar contribution to the SH field for an isolated nanosphere might be small, leading to a quadrupolar radiation pattern. However, in a spatially structured system, the polarizing local field may have small lengthscales. For example, a sphere lying close to a surface is subject to its own image field, which would have a lengthscale of variation of the order of its distance to the surface. In Fig. 13 we show the $p \to p$ SH radiation patterns for a nanosphere at different distances from an underlying substrate.[33] For

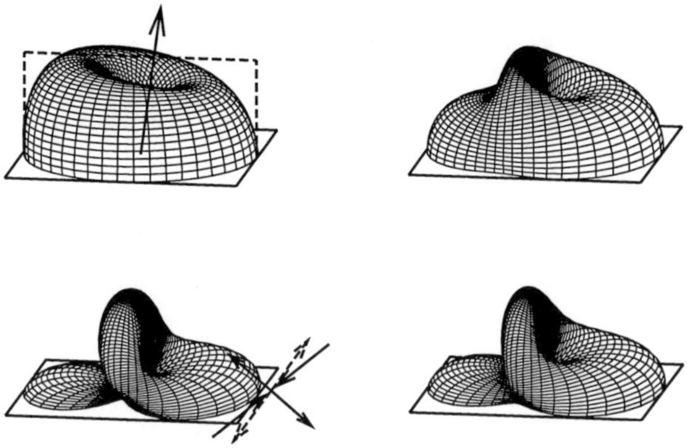

Figure 13. Angular distribution of the SHG from a nanosphere above a substrate at distances that vary from its radius R to twice its radius (clockwise from top left). The dipolar axis at the closest distance and the quadrupolar principal axes at the largest distance, as well as the plane of incidence, are indicated.[33]

the closest distance the pattern is dominated by the dipolar radiation which gradually evolves into quadrupolar radiation with tilted principal directions as the distance is increased.

Going back to SH by nanoparticle composite films, we remark that as the field gradient for a plane wave points along its propagation direction, and as the field is perpendicular to the propagation direction, neither the SH dipole nor the SH quadrupole induced in a nanosphere is capable of radiating along the forward radiation $\theta = 0°$, in apparent contradiction to the experimental results. However, SH experiments are typically done with tightly focused beams, so that there is an additional *lateral gradient* which points from the edge towards the center of the beam (Fig. 14). The gradient along the transverse direction may tilt the induced dipoles so that they may radiate at small angles, and along the forward direction itself. However, the fields produced by dipoles on opposite sides of the beam axis cancel each other at exactly $\theta = 0°$. On the other hand, the coherent sum of the fields produced by all the illuminated particles interfere destructively for large angles, so that the SH field may only be produced within a narrow cone whose angular divergence is similar to the divergence $\theta_1 \sim \lambda/w_0$ of the fundamental beam, where w_0 is the waist of the beam. Therefore, we expect a two-lobed intensity distribution (Fig. 15a). To obtain Fig. 15a,

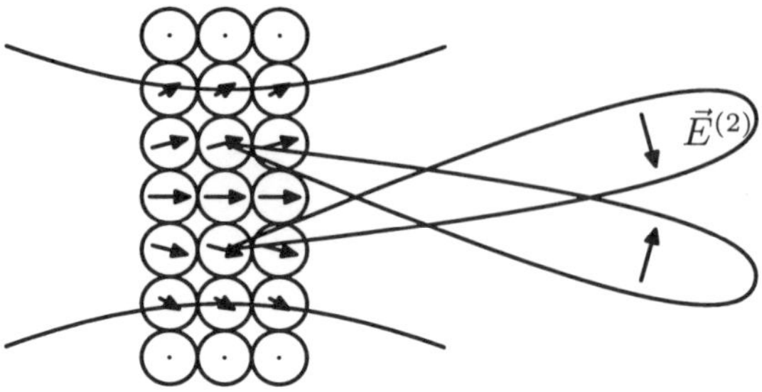

Figure 14. Nanoparticle composite film illuminated by a finite waisted beam. The gradient along the transverse direction tilts the SH dipoles, so that they may radiate close to the forward radiation.

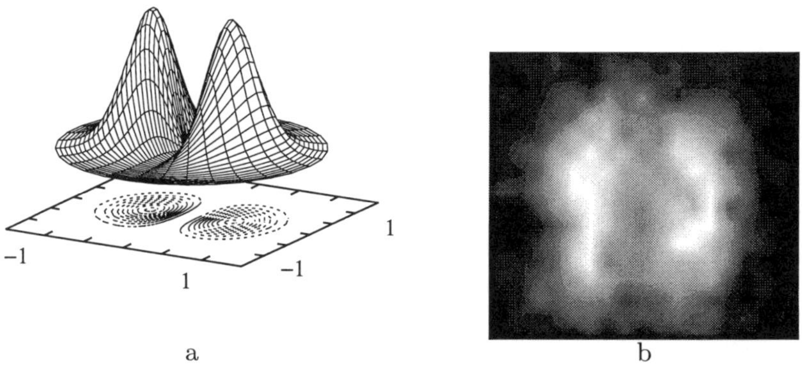

Figure 15. Angular distribution of the SH radiation from a thin composite film illuminated by a Gaussian beam linearly polarized along the line that joins the two peaks. (a)Theoretical prediction.[34] The angles are measured in units of the angular divergence of the fundamental beam. (b)Experimental result.[36]

the macroscopic SH polarization

$$\vec{P}^{(2)} = \Gamma \nabla E^2 + \Delta' \vec{E} \cdot \nabla \vec{E} \tag{39}$$

was obtained from Eq. (25) substituting Eqs. (37) and (38), which therefore yield the macroscopic nonlinear susceptibilities of the composite[34]

$$\Gamma = \frac{n}{18}(9\gamma^m + \gamma^q - 3\tilde{\gamma}^q), \tag{40}$$

and

$$\Delta' \equiv n(\gamma^e - \gamma^m - \gamma^q/6), \tag{41}$$

where n is the number density of the nanoparticles. From $\vec{P}^{(2)}$, the SH transverse current, radiated vector potential, electromagnetic fields, Poynting vector and angular distribution of the intensity are calculated following the usual electrodynamic approach.[34] The result given by Fig. 15a has been confirmed experimentally (Fig. 15b).[36] In the former experiments,[31] one of the peaks was missed and the other was assumed to correspond to the forward direction. Richer, many-lobed patterns, including the possibility of true forward SH scattering have been predicted for input beams of higher order gaussian profiles and for beams with a superposition of gaussian modes.[37]

One of the curious consequences of the theory above is that within the composite, and more generally, within any centrosymmetric film illuminated by a finite beam, there is a SH polarization $\vec{P}^{(2)}$ capable of radiating. However, it is of a multipolar character and is therefore not proportional to $\vec{E}\vec{E}$ but to $\vec{E}\nabla\vec{E} \sim E^2/w_0$. Thus, the total SH efficiency from a thin film of width l,

$$\mathcal{R} \equiv \frac{\mathcal{P}^{(2)}}{\mathcal{P}^2} = \frac{64\pi^2}{c}\frac{(ql)^2}{w_0^4}|\Delta'|^2, \tag{42}$$

is inversely proportional to w_0^4. This means that the SH power $\mathcal{P}^{(2)}$ is not proportional to the square of the input power $\mathcal{P}$, but to the square of the input intensity I. As an increase of the input power is usually accompanied by an increase of the waist, in order to avoid damaging the sample, it may happen paradoxically that the SH efficiency and the SH output power decrease as the input power increases.

Very recently,[36,38] a solution to increase the multipolar SHG from centrosymmetric systems has been proposed and succesfully tested. By splitting the fundamental beam into two equal beams which cross each other within the sample at a finite angle, an interference pattern develops. The spatial scale of variation of this pattern is of the order of λ instead of w_0. Therefore, the SH signal may be enhanced several orders of magnitude by an amount $\sim (w_0/\lambda)^2$. Curiously, this enhancement is maximum when the two beams have perpendicular polarizations and therefore interference produces no intensity oscillations, although it does produces a fast spatial modulation of the total polarization. On the other hand, the enhancement disappears when both polarizations are parallel.

11. Summary

We have discussed the origin of the surface and bulk second harmonic generation from centrosymmetric systems, and we have illustrated its calculation using a few exactly solvable models. We have presented both phenomenological and microscopic aspects of the problem and we have presented a few applications, among which we included the prediction of a huge enhancement of the SHG signal at the multipolar surface plasmon resonance of simple metals, recently observed and reported in an accompanying lecture, and the radiation patterns of single nanoparticles and nanocomposite materials. SHG has been firmly established experimentally as a very useful probe of surfaces and interfaces, but its understanding and interpretation through theory is still evolving. An example of the interplay of theory and experiment was the realization that a two beam geometry may enhance the SH signal by several orders of magnitude.

Acknowledgements

This work was partially supported by DGAPA-UNAM under grant No. IN117402.

References

1. Robert W. Boyd, *Nonlinear Optics*, 2nd. Ed. (Academic Press,New York,2003).
2. T.F. Heinz, in *Nonlinear Surface Electromagnetic Phenomena*, Ed. by H.E. Ponat and G.I. Stegeman (Elsevier Science, Amsterdam, 1991); Chap. 5.
3. G.A. Reider, T.F. Heinz, in *Photonic Probes of Surfaces*, Ed. by P. Halevi (Elsevier Science, Amsterdam, 1995); Chap. 9.
4. J. F. McGilp, Surf. Rev. and Lett. **6**, 529 (1999).
5. *Photonic Probes of Surfaces* Ed. by P. Halevi (Elsevier Science, Amsterdam, 1995).
6. A. Cricenti, J. Phys. Cond. Matter **16** S4243 (2004).
7. Yu.A. Il'inskii, L.V. Keldysh, *Electromagnetic Response of Material Media*, (Plenum Press, New York, 1994).
8. P.N. Butcher, D. Cotter, *The Elements of Nonlinear Optics* (Cambridge University Press, 1990).
9. E. Adler, Phys. Rev. **134**, A728 (1964).
10. N. Bloembergen, R.K. Chang, S.S. Jha, C.H. Lee, Phys. Rev. **174** 813 (1968); *ibid.* **178** 1528(E) (1969).
11. J.F. Nye *The Physical Properties of Crystals* (Clarendon Press, Oxford, 1985).
12. Sipe et al., Phys. Rev. B **35**, 1129 (1987).
13. A.V. Petukhov, A. Liebsch, Surf. Sci. **334** 195 (1995).

14. A. Liebsch, *Electronic Excitations at Metal Surfaces*, (Plenum Press, New York, 1997).
15. J. Rudnick, E.A. Stern, Phys. Rev. B **4** 4274 (1971).
16. P. Guyot-Sionnest, A. Tadjeddine, A. Liebsch, Phys. Rev. Lett. **64** 1678 (1990).
17. P. Guyot-Sionnest, W. Chen, Y.R. Shen, Phys. Rev. B **33** 8254 (1986).
18. A.V. Petukhov, A. Liebsch, Surf. Sci. **294** 381 (1993).
19. Bernardo S. Mendoza, W. Luis Mochán, Phys. Rev. B **53** 4999 (1996);*ibid.* **61** (E)16243 (2000).
20. J.D. Jackson, *Classical Electrodynamics*, 2nd. ed. (Wiley, New York, 1975).
21. M. Corvi and W. L. Schaich, Phys. Rev. B **33**, 3688 (1986).
22. V. Mizrahi and J. E. Sipe, J. Opt. Soc. Am. B **5**, 660 (1988).
23. Bernardo S. Mendoza, W. Luis Mochán, Phys. Rev. B **55** 1 (1997); *ibid.* **53** R10473 (1996).
24. J.A. Maytorena, W. L. Mochán, B.S. Mendoza, Phys. Rev. B **51** 2556 (1995).
25. N.D. Lang, W. Kohn, Phys. Rev. B **1**, 4555 (1970).
26. G. Chiarello, V. Formoso, A. Santaniello, E. Colavita, L. Papagno, Phys. Rev. B **62** 12676 (2000).
27. K.-D. tsuei, E.W. Plummer, A. Liebsch, K. Kempa, P. Bakshi, Phys. Rev. Lett. **64** 44 (1990); K.-D. tsuei, E.W. Plummer, A. Liebsch, E. Pehlke, K. Kempa, P. Bakshi, Surf. Sci. **247** 302 (1991).
28. S.R. Barman, H. Haberle, K. Horn, Phys. Rev. B **58**, R4285 (1998)
29. K. Pedersen, T.G. Pedersen, P. Morgen, in this volume.
30. R.J. Cole, D.S. Roseburgh, J. Phys. Condens. Matter **16** S4279 (2004).
31. Y. Jiang et al., Appl. Phys. Lett. **78**, 766 (2001).
32. J.I. Dadap, J. Shan, K.B. Eisenthal, T.F. Heinz **83** 4045 (1999).
33. Vera L. Brudny, Bernardo S. Mendoza, W. Luis Mochán, Phys. Rev. B **62** 11152 (2000).
34. W. Luis Mochán, Jesús A. Maytorena, Bernardo S. Mendoza, Vera L. Brudny, Phys. Rev. B **68** 085318 (2003).
35. Jerry I. Dadap, Jie Shan, Tony F. Heinz, J. Opt. Soc. Am. B **21** 1328 (2004).
36. P. Figliozzi et al., Phys. Rev. Lett. **94**, 047401 (2005).
37. Roberto Bernal, Jesús A. Maytorena, Phys. Rev. B **70** 125420 (2004).
38. Liangfeng Sun et al. (to appear in Optics Letters).

EXCITATION OF MULTIPOLE PLASMON IN OPTICAL SECOND-HARMONIC GENERATION

K. Pedersen[1], T. G. Pedersen[1], P. Morgen[2]

[1]*Department of Physics and Nanotechnology, Aalborg University, Aalborg Øst, Denmark*
[2]*Fysisk Institut, University of Southern Denmark-Odense University, Denmark*

The second-order nonlinear optical response of thin Cs and K films has been investigated by optical second-harmonic generation spectroscopy (SHG). Deposition of Cs on Si(111)7×7 at low temperature leads to orders of magnitude increase of the SH signal compared to that of the clean surface as the first monolayer of alkali metal is adsorbed. Spectra recorded for more than approximately 2 layers of Cs are completely dominated by resonance peaks at 2.25 eV and 4.5 eV corresponding to one photon and two photon excitations of multipole plasmons appearing at $0.8\omega_P$. Spectra recorded for lower Cs coverage are shifted to lower energies, indicating an effective electron density lower than that of bulk Cs. Cs films deposited on a Ag(111) surface formed on the Si surface show spectra that are essentially identical to those of Cs on Si. The multipole plasmon is thus a collective mode that is intrinsic to the alkali metal surface. Deposition of K on Si leads to spectra with a double peak at 2.85 and 3.15 eV and a smaller peak at the E_2 critical point at 4.5 eV. The double peak corresponds to surface and multipole plasmons, respectively in the K film indicating that the K film is rough. When K is deposited on a Ag(111) film grown on Si(111) a smooth alkali film showing only the multipole resonance is formed.

1. Introduction

Excitation of plasmons is of fundamental importance in the optical response of metal surfaces [1,2]. Surface plasmons have been studied extensively over the years both for fundamental reasons and for applications where the confinement and enhancement of optical fields at surfaces can generate interesting new sensor concepts or integrated optical circuits [3]. Excitations of surface plasmons on rough metal surfaces leads to local-field enhancement that can generate orders of magnitude higher SHG than the corresponding flat surfaces [4]. Also SHG enhancement in periodic gratings has been studied [5]. For atomically clean surfaces a so-called multipole plasmon mode (at frequency ω_M) has been found between the bulk (ω_P) and the surface (ω_S) plasmon modes. The multipole plasmon was first predicted by Bennet [6] and has later been described theoretically in details through density functional calculations [2]. It has been observed directly in electron energy loss spectroscopy experiments for several metals [7]. The oscillations in the tail of the charge distribution in vacuum leaves the center of the distribution fixed, contrary to surface plasmon oscillations. In the multipole eigenmode the low-density part of the density tail moves outward while the high-density part moves toward the surface, which has lead to the term 'multipole' for the characterization of this mode.

Density functional calculations of the SHG spectra for alkali metal overlayers on Al show that the nonlinear response of the surface is completely dominated by a resonance at the multipole plasmon frequency [8]. At resonance the calculations predict that a single monolayer of Cs or K leads to about three orders of magnitude enhancement of the SH signal compared to that of a clean Al surface. Several experiments have demonstrated that SHG is highly sensitive to the presence of alkali metal atoms on the surface [9-11]. When the overlap of the adatom wave functions is small at low coverage the contribution to SHG could come from excitation of alkali adatom levels that are broadened by coupling to the conduction band of the substrate. At higher coverage the metallic properties of the alkali metal atoms dominate and may lead to orders of magnitude enhancement of SHG [9,10]. Previous SHG experiments, being based on only a few pump wavelengths, have not been able to distinguish between the different possibilities for the plasmon resonance frequency. In the present spectroscopic SHG experiments it is demonstrated that the multipole surface resonance dominate the SH spectra of smooth alkali metal surfaces.

2. Experimental details

The experiments were performed on samples mounted in a vacuum system equipped with low energy electron diffraction (LEED) and a manipulator allowing cooling to 150K and direct resistive heating of the samples. Sample cleanliness was verified by Auger electron spectroscopy in combination with LEED. The samples were cut from a 1-mm thick *n*-type wafer with a resistivity of 5 Ω-cm. Alkali metals were deposited from well outgased dispensers. The SHG experiments were performed with an optical parametric oscillator pumped by the third harmonic of 6-ns pulses from a Q-switched Nd:YAG laser. All data were recorded with 60° angle of incidence and both the fundamental and the detected SH light were polarized in the plane of incidence (p-polarized). After filtering with coloured glass filters the SH signals were detected by a photo multiplier tube connected to a boxcar integrator. The SH signals were normalized against those generated in reflection from the input side of a wedge shaped quartz crystal.

3. Experimental results and discussion

Deposition of alkali metals on Si(111)7×7 at room temperature leads to a dramatic drop in the sticking probability when the primary adsorption sites on the substrate are occupied [12]. At lower temperatures, however, a high sticking probability is maintained and multiplayer films may be formed. In order to follow the growth process with SHG a wedge shaped Cs film was grown over a length of 15 mm on the Si(111)7×7 surface at 150 K by slowly moving the sample into the shadow of a fixed shield during the deposition. Subsequently, the thickness dependence of SHG was recorded by scanning the Cs wedge through a focused laser beam. Figure 1 shows such scans recorded at different photon energies. The SH signals, shown as a function of deposition time, have been normalized to the same height. The relative signal levels at different photon energies can be seen in Fig. 2. All scans in Fig. 1 show relatively fast increase in SH signal after 10 min deposition and a constant level after 17 min deposition. Apart from large differences in signal level at the thick end of the film the major difference between different photon energies is in the submonolayer behaviour.

Figure 2 shows the SHG signal as a function of the SH photon energy recorded after Cs deposition for respectively 15 and 60 min. Both spectra show two peaks that could correspond to one-photon and two-photon excitation of a resonance. The resonance shifts to higher energy with deposition time. After 30 min. deposition no change in the spectrum was observed upon further deposition. The saturation positions of the peaks at 2.25 and 4.5 eV correspond to, respectively $0.8\omega_P$ and $1.6\omega_P$, where ω_P=2.8 eV is the bulk plasma frequency of Cs. The peak at 4.5 eV falls at the position of the E_2 critical point, but the signal level is more than an order of magnitude higher than that of the E_2-resonance of the clean 7×7 surface. This together with the absence of the peak in the spectrum recorded at lower coverage points at a two-photon plasmon resoanance as the origin of the 4.5-eV peak. It is interesting to note that there is no sign of resonances at the bulk plasma frequency or the surface plasma frequency at ω_S=2.0 eV. The spectra also indicate a resonance at the E_0/E_1 critical point. The spectrum for 15 min. deposition shows a sharp hole at 3.5 eV that could be due to destructive interference between the plasmon contribution and the contribution from the Si/metal interface. Unfortunately this falls in the range from 730 to 690 nm around the degeneracy of the OPO.

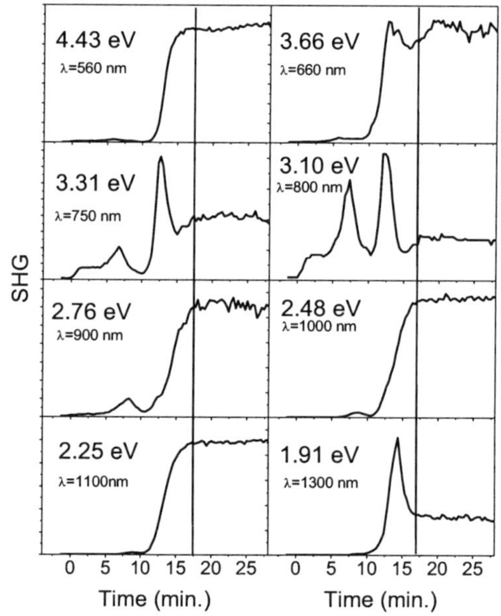

Figure 1. SHG as a function of the thickness of a Cs film deposited on Si(111)7×7 at 150 K recorded at different photon energies. The energies specified are the SH photon energies while the wavelengths are for the pump beam. The vertical line marks the coverage where the transition to a plasmon dominated signal has taken place.

With the help of the spectra in Fig. 2 it is now possible to understand the growth curves shown in Fig. 1. When either the fundamental or the SH frequency is at resonance (2.25 or 4.5 eV) a monotonous increase in the SH signal is recorded because the system moves towards the highest saturation signal for a thick film. Between the one photon and two photon resonances (3.0 – 4.0 eV) the shift of the resonances with Cs coverage leads to local maxima close to monolayer coverage. This is also the case when the SH photon energy is just below the resonance (1.91 eV). At lower coverage the scans at 3.3 and 3.1 eV show additional maxima after 5-10 min deposition. The same maxima can be found in the rest of the scans when the part below monolayer coverage is scaled up. In this submonolayer regime it is expected that excitation of hybridised Cs atomic levels dominate the adsorbat-induced SH signal. It is suggested that these maxima are related to saturation of different adsorption sites for Cs atoms in the 7×7 structure.

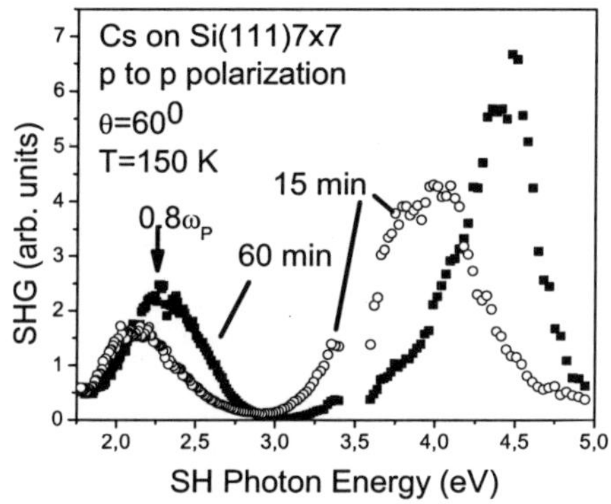

Figure 2. SHG as a function of the SH photon energy recorded for 15 and 60 min of Cs deposition on a Si(111)7×7 at 150 K.

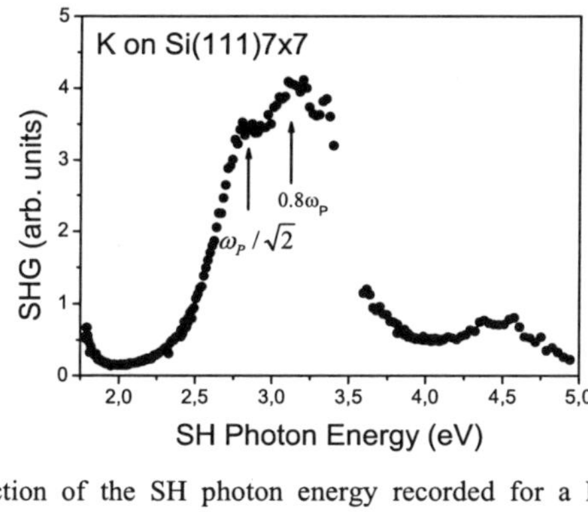

Figure 3. SHG as a function of the SH photon energy recorded for a K film deposited at a Si(111)7×7 at 150 K.

Deposition of K on the Si(111)7×7 surface under the same conditions as those used for Cs leads to a broad peak around 3 eV as shown in Fig. 3. A second peak at 4.5 eV coincides with the E_2 critical point of Si and is thus caused by excitations at the Si/K interface. The corresponding peak from the E_0/E_1 critical point (around 3.4 eV for the clean surface) is difficult to observe since it falls in the range of the large K-induced peak around 3 eV. The major peak consists of two contributions at 2.8 eV and 3.1 eV. With a bulk plasma frequency of 3.8 eV these two contributions correspond to the surface plasmon and multipole plasmon frequencies, respectively. Thus, contrary to the Cs film the K film shows a surface plasmon mode.

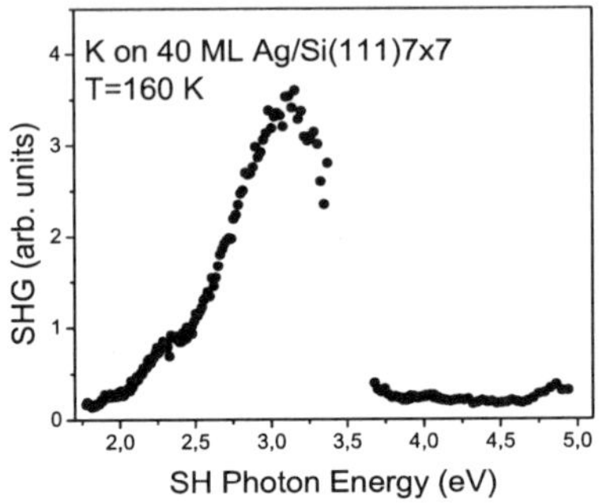

Figure 4. SHG as a function of the SH photon energy recorded for a K film deposited on a 40-ML Ag(111) film on Si(111)7×7 at 150 K.

In order to investigate the film growth further a 40-ML Ag film was grown on the clean 7×7 surface at 150 K. This is known to lead to a film consisting of small domains with curved top surfaces. After room temperature annealing for a few hours the small domains transform into larger (111)-oriented islands with atomically flat surfaces. Such Ag surfaces have been used as substrates for growth of alkali metal films. Both Cs and K deposition on the Ag surface lead to a spectra with resonances at the respective multipole plasmon frequencies. Thus, the surface plasmon is found when K is deposited on the Si surface, but not for K on the Ag surface. Since it is well known that the surface plasmon is sensitive to the structure of the surface on the atomic monolayer level [13] this demonstrates that the K film is more rough than the Cs film grown under the same conditions. Deposition of K on Si(111) thus gives rise to substantial island growth after formation of the first closed layer of K in the same way as it was concluded from core-level spectroscopy on the K/Si(100)c(4×4) system by Ghao et al. [14].

4. Summary

SHG spectroscopy on thin Cs and K films show that the multipole plasmon resonance dominates the surface SH response of these films. The multipole plasmon resonance is found for alkali metal films deposited both on Si(111)7×7 and on Ag(111) films grown on Si. The resonance is fully developed after deposition of the first few atomic layers, which demonstrates that the mode is localized to the outermost part of the surface density profile. This surface localization of the mode combined with the surface selectivity of second order nonlinear optical processes makes the multipole plasmon the dominating excitation mechanism in SHG from flat alkali metal surfaces.

References

1. F. Forstmann and R. R. Gerhardts, *Metal Optics Near the Plasma Frequency,* Springer Tracts of Modern Physics **109** (Springer, Berlin, 1986).
2. A. Liebsch, Phys. Rev. Lett. **67**, 2858 (1991).

3. S. I. Bozhevolnyi and Valentyn S. Volkov, Appl. Phys. Lett. **79**, 1076 (2001).
4. G. T. Boyd, Th. Rasing, J. R. R. Leite, and Y. R. Shen, Phys. Rev. B **30**, 519 (1984).
5. J. L. Coutaz, M. Nivière, E. Pic, R. Reinisch, Phys. Rev. B **32**, 2227 (1985).
6. A. J. Bennet, Phys. Rev. B **1**, 203 (1970).
7. K.-D. Tsuei, E. W. Plummer, A. Liebsch, K. Kempa, and P. Bakshi, Phys. Rev. Lett. **64**, 44-47 (1990).
8. A. Liebsch, Phys. Rev. B **40**, 3421-24 (1989).
9. H. W. K. Tom, C. M. Mate, X. D. Zhu, J. E. Crowell, Y. R. Shen, and G. A Somorjai, Surf. Sci. 172, 466 (1986).
10. K. Pedersen, P.-E. Hansen, and P. Morgen, phys. Stat. Sol (a) 170, 411 (1998).
11. J. Boness, G. Marowsky, J. Braun, G. Witte, and H.-G. Rubahn, Surf. Sci. **402-404**, 513 (1998).
12. K. O. Magnusson and B. Reihl, Phys. Rev. B 41, 12071 (1990).
13. J. G. Endriz and W. E. Spicer, Phys. Rev. B **4**, 4144 (1971).
14. Y.-C. Chao, L. S. O. Johansson, and R. I. G. Uhrberg, Surf. Sci. 372, 64 (1997).

NON-LINEAR OPTICAL PROBES OF BIOLOGICAL SURFACES

MISCHA BONN

FOM Institute for Atomic and Molecular Physics, Kruislaan 407, 1098 SJ Amsterdam, The Netherlands; Leiden Institute of Chemistry, Leiden University, P.O. Box 9502, 2300 RA, Leiden, The Netherlands.

VOLKER KNECHT

Department of Biophysical Chemistry, University of Groningen, Nijenborgh 4, 9747 AG Groningen, The Netherlands

MICHIEL MÜLLER

Swammerdam Institute for Life Sciences, University of Amsterdam, P.O. Box 94062, 1090 GB Amsterdam, The Netherlands

We introduce the application of non-linear surface probes to investigate intermolecular interactions between different molecules constituting a (model system) membrane. As opposed to many of the linear spectroscopic tools that are applied to these systems, non-linear techniques provide direct information on molecular conformation, orientation and order without the need for markers, such as fluorescent probes. These capabilities are illustrated with examples of second-order non-linear sum-frequency generation (SFG) and third-order non-linear coherent anti-Stokes Raman scattering (CARS) microscopy. These techniques have the additional advantage that they are sensitive to molecular vibrations, which provide an intrinsic molecular probe for chemical identity, structure, intra- and intermolecular interactions and dynamics. We also present Molecular Dynamics simulations that support the experiments.

1. Introduction

1.1. *Intermolecular interactions in membranes*

Of the many surfaces known to mankind, biological surfaces, particularly membranes, are arguably the most complex. The various cellular membranes (plasma membrane, nuclear membrane, internal organelles) all show a significantly different lipid composition and, moreover, different lipid species are distributed asymmetrically over the two leaflets that constitute the lipid bilayer. The membrane complexity appears on many levels: First of all, the membrane composition is very heterogeneous: the variety and number of molecular constituents is astounding. To name the variation in the three most important constituents:

- Lipids: lipids form the basic building block of cell membranes. Owing to their bipolar nature (lipids consist of a polar head group and a long apolar tail), they can self-organize and thus form the skeleton of living cells. Over a thousand different lipids have been found in membranes.
- Proteins: proteins constitute up to tens of percents of the membrane. Proteins are the membrane machinery and regulate, amongst others, the flow of material and information in- and out of the cell.
- Sterols: sterols –mainly cholesterol for mammalian cells– constitute up to ~40 % of membrane material and are the key means by which cells modulate membrane properties such as permeability and protein action. They do so by affecting the phase and stiffness of the lipid layer.

Secondly, these different molecular constituents are thought to be inhomogeneously, non-randomly distributed over the surface, presumably in distinct functional domains with different thermodynamic and elastic properties. This means that the surface is mesoscopically inhomogeneous.

Thirdly, the complexity also is expressed on a molecular level in the constituents themselves; not only is there a strong variation in e.g. lipid types, but for one specific lipid, many molecular conformations are possible. The apolar acyl chain, for example, may be perfectly straight (all-trans), but it can also be strongly bent (gauche kinks), depending on its local environment.

The complexity of the membrane surface is such that its basic structure remained elusive until the 1970's: it was not until then that it was recognized that membrane lipids are organized into bilayers [1]. In the subsequent decades, it has become apparent that the membrane complexity is intimately related to its functionality. Very recently, for example, there has been substantial debate on the possibility that domain formation plays a key role in the functioning of the membrane [2]. There is evidence that so-called 'lipid rafts' are formed; these are agglomerations of specific lipids in the membrane in a different thermodynamic state than the bulk of the membrane. Specific proteins can cluster in the raft, and putative biological function of lipid rafts includes: intracellular lipid and protein sorting and transport; transmembrane signaling; membrane docking and membrane fusion; and, very importantly, viral structural proteins, including those of HIV-1 Sendai, measles and influenza virus, reportedly associate with lipid rafts. One key player in the formation of lipid rafts is cholesterol; Cholesterol has been demonstrated to be an essential ingredient for the formation of lipid rafts.

Despite its apparent importance, little is known regarding the interaction of cholesterol with bio-lipids on a molecular level, and how, as a result of that interaction, cholesterol affects the phase behavior of the lipids.

It is the aim of this contribution to demonstrate that non-linear optical techniques are very well suited to investigate the phase behavior of phospholipids in model systems for biological membranes. The strength of these techniques is that they provide a way to investigate the molecular properties of membranes in a non-invasive, non-destructive manner that does not require the use of probe molecules, such as fluorescent labels, that may alter the properties of the membranes.

1.2. *Non-linear optics*

The field of non-linear optics is concerned with all light-matter interactions in which the response of the material depends in a non-linear fashion on the input electromagnetic field strength. Excellent text books exist (e.g. [3-6]) that provide both a general overview of the field and a detailed theoretical analysis. Here some of the main features are reviewed which are relevant for the application of non-linear optical techniques to surface spectroscopy. The discussion is limited to so-called parametric processes, which are processes for which the initial and final quantum-mechanical states of the system are equal. These processes include phenomena like sum-frequency generation and stimulated Raman scattering.

The standard approach in non-linear optics is to split the macroscopic polarization P - i.e. the result of an ensemble of oscillating dipole moments - in a linear and a non-linear part and to expand it in successive orders of the electromagnetic field (E):

$$P(r,t) = P_L(r,t) + P_{NL}(r,t) = \chi_L^{(1)} \cdot E + \chi_{NL}^{(2)} \cdot EE + \chi_{NL}^{(3)} \cdot EEE + \dots \qquad (1)$$

In equation (1), $\chi_L^{(1)}$ and $\chi_{NL}^{(n)}$ denote the linear and non-linear susceptibilities, respectively, which relate the complex amplitude of the electromagnetic fields to the polarisation. In general, for a material with dispersion and/or loss, the susceptibility is a complex tensor quantity. Note that from symmetry arguments it follows immediately that even orders of the non-linear susceptibility vanish for materials that possess a centre of inversion symmetry. As such, even-order non-linear techniques are selectively surface sensitive. And have the additional advantage over conventional IR spectroscopy that it is a background-free technique. We will use these properties to selectively study surfaces using sum-frequency generation (SFG). Various approaches have been developed to relate the macroscopic susceptibility to microscopic quantities (see e.g. [7, 8]).

The generated macroscopic non-linear polarization P_{NL} in Eq. (1) acts as a source term in the wave equation, producing an electromagnetic signal wave. Accordingly, the recorded signal intensity is proportional to the complex square of the field.

As an example we consider the process of sum-frequency generation (SFG) [9], schematically depicted in the left panel of Fig. 1. Two electromagnetic waves of frequencies ω_{IR} and ω_{VIS} interact with the material to produce a signal

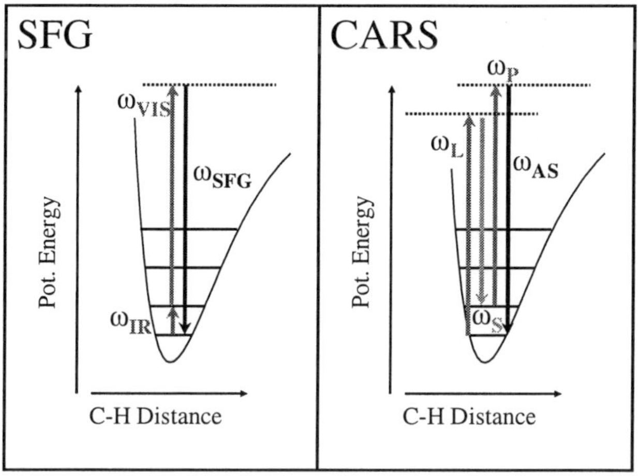

Fig. 1. Energy level diagrams for the parametric non-linear optical processes of (a) SFG and (b) CARS. Common to both is that they are electronically non-resonant but vibrationally resonant. The various laser and signal frequencies represent: infrared (IR), visible (VIS), sum-frequency (SFG) for the case of SFG, and Laser (L), Stokes (S), Probe (P) and Anti-Stokes (AS).

wave at the sum frequency. Thus the SFG intensity is related to the second-order susceptibility through:

$$I(\omega_{SFG}) \propto |P(\omega_{SFG})|^2; \; P(\omega_{SFG}) \propto \chi^{(2)}(\omega_{SFG} = \omega_{IR} + \omega_{VIS}) \cdot E(\omega_{IR}) \; E(\omega_{VIS}) \quad (2)$$

where the explicit spatial and temporal dependence of the electromagnetic fields has been dropped.

The right-hand side of Fig. 1 depicts the third-order non-linear process of Coherent Anti-Stokes Raman Scattering (CARS). CARS is the non-linear optical analogue of spontaneous Raman scattering. While amplifying the Raman signal by many orders of magnitude and enhancing the detectability of the signal against background luminescence, it retains the unique spectral specificity of Raman scattering. The Raman spectrum can provide a molecular 'fingerprint' even at room temperature and in complex environments such as living cells, through the vibrations of molecules in the sample. As such, it is also very sensitive to intermolecular interactions. Recent developments in laser technology have permitted the introduction of this established non-linear spectroscopic technique into the field of high resolution microscopy [10,11]. As CARS spectroscopy relies on visible (short wavelength) rather than IR radiation to probe the vibrational response, the two different laser beams (interacting with the samples once and twice, respectively) can be focused down to sub-micron length scales, allowing for a spatial resolution of some hundreds of nanometers.

Note that as a third-order non-linear optical process used for microscopy, the origin of the CARS signal is not restricted to the surface, but may results from the bulk of the material contained within the focal volume. More particularly, for CARS the induced macroscopic polarisation is given by:

$$P(\omega_{AS}) \propto \chi^{(3)}(\omega_{AS} = \omega_L - \omega_S + \omega_P) \cdot E(\omega_L)E(\omega_S)E(\omega_P) \qquad (10)$$

where ω_L, ω_S, ω_P and ω_{AS} denote the frequencies of the so-called *Laser*, *Stokes*, *Probe* and *Anti-Stokes* fields respectively.

2. Sum-frequency generation spectroscopy of lipid monolayers

Lipid molecules, such as the phospholipid L-1,2-Dipalmitoyl-sn-glycero-3-phosphocholine (DPPC) depicted in the left of Fig 2, consist of a polar headgroup and two long apolar alkyl chains. This bipolar structure gives rise to fascinating behavior when put in contact with water, such as self-organization and separation into different phases. Monolayers are excellent model systems for membrane biophysics, since a biological membrane (bilayer) can be considered as two weakly coupled monolayers. Of particular interest is the phase behavior as a function of phospholipids density, which has been studied with various thermodynamic and spectroscopic techniques [12]. This interest originates from the realization that the molecular and microscopic structure associated with the density dependent phase behavior of phospholipid layers is essential for a complete understanding of cell membrane and vesicle properties.

Using SFG, we analyze the order and orientation of the alkyl chains as a function of density within a monolayer [13]. The density is varied in a so-called Langmuir trough, depicted schematically in Fig. 1. The surface density of phospholipids can be varied simply by movement of a Teflon barrier. We record the surface tension, fluorescence image and SFG spectrum under identical conditions. The results of the latter are depicted in the central panel of Fig. 2.

For the strongly compressed DPPC layer (47 Å^2/molecule), only the resonances of the terminal CH_3-groups are observed (symmetric and asymmetric CH_3 stretch vibrations and a Fermi resonance). Remarkably, resonances associated with the methylene (CH_2) groups are absent, despite the fact that these are present in a 15 fold higher density than the CH_3-groups. This is a direct consequence of the selection rules of SFG: SFG is inherently sensitive only to media lacking inversion symmetry. The alkyl strands of the DPPC molecule consist of an even number of CH_2 groups. If the alkyl chain is fully stretched and in the (zigzag) all-trans conformation, each C-C bond has a local

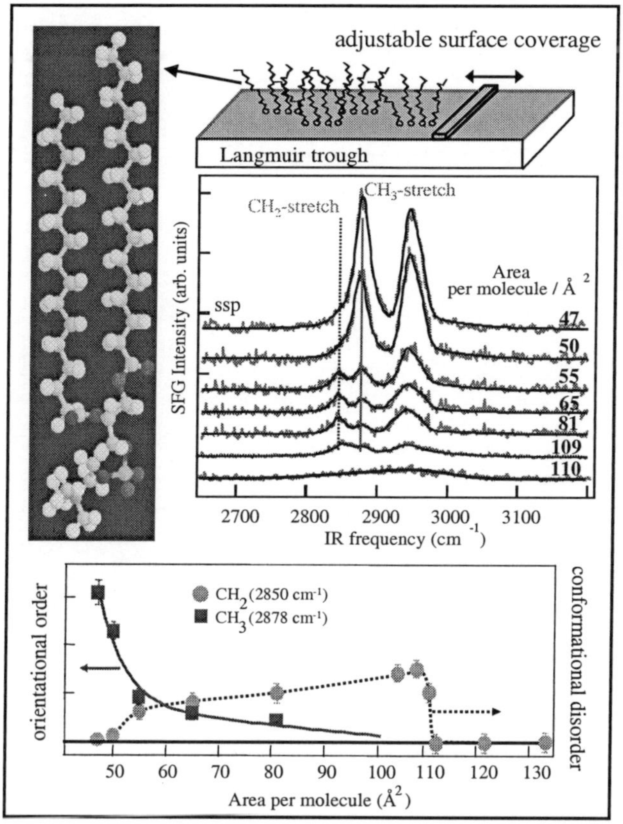

Fig. 2. SFG spectroscopy on phospholipids monolayers summarized: The top figure shows a Langmuir trough used in the experimentsEnergy level diagrams for the parametric non-linear optical processes of (a) SFG and (b) CARS. Common to both is that they are electronically non-resonant but vibrationally resonant. The various laser and signal frequencies represent: infrared (IR), visible (VIS), sum-frequency (SFG) for the case of SFG, and Laser (L), Stokes (S), Probe (P) and Anti-Stokes (AS).

center of inversion symmetry. The absence of signal thus establishes that the compressed DPPC layer consists of fully stretched alkyl chains.

Upon decompressing the layer, CH_2-resonances indeed appear (see Fig. 2, at 55 Å^2/molecule) due to the disorder that appears within the molecules: as the phospholipids have more space, the apolar tails becomes disordered, and the inversion symmetry that was present in the chains at high compression is lifted. Simultaneously, the CH_3 signal decreases rapidly, with increasing molecular area. This is a consequence of the orientational disorder that appears upon decompression: The polarization that radiates the SFG signal is largest when all the CH_3 dipoles are collectively aligned; upon decompression, this alignment vanishes.

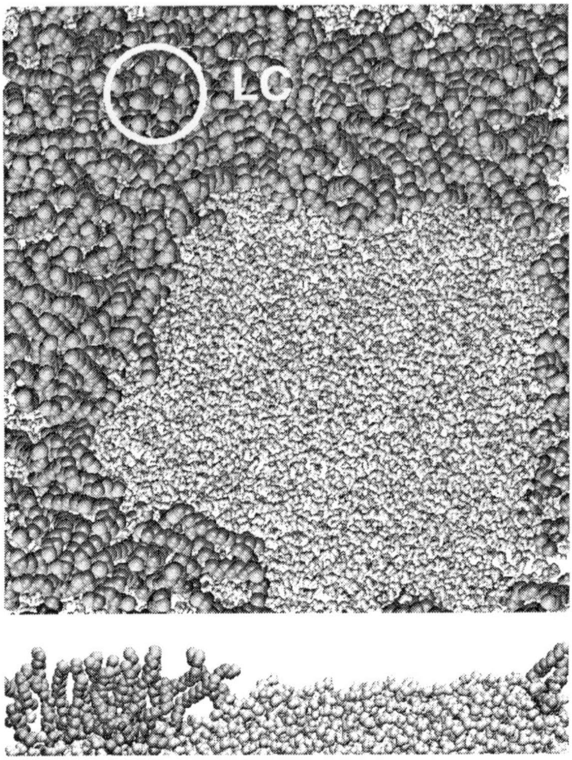

Fig. 3. Snapshots of large DPPC monolayer patches at 114 Å^2 per molecule after 21 ns of molecular dynamics simulation, viewed from the air phase. Note that the DPPC monolayer has ruptured, exposing a large amount of water (light colour). The lower panel depicts a slice along the diagonal from the upper left to the lower right. The white circle marks a domain of lipids in the liquid compressed phase, indicated by a hexagonal arrangement of terminal CH_3 groups of the lipid tails.

These results are depicted schematically in Fig. 2: As the orientational order is diminished (CH_3 signal), the conformational, molecular disorder in the phospholipids (CH_2 signal) increases.

A surprising observation is made at a density between 109 and 110 Å^2/molecule [13]: the resonant signal vanishes very suddenly. The signal at very low densities (>110 Å^2/molecule) originates mostly from the CH_2 groups, which became visible because the inversion symmetry was lifted within the layer upon decompression. The transition indicates that the lifted inversion symmetry reappears at high surface area per molecule. Therefore, this transition is indicative of one of two things: either the order with the alkyl chains is restored [14], or the transition is into a state with complete disorder [15]; in either of these two scenarios, inversion symmetry is restored, accounting for the disappearance of the signal. Initially, the transition was ascribed to a 'curling up' of the phospholipids molecules. This state could be energetically favorable, as it would reduce the exposure of the hydrophobic alkyl chain to the polar water. The complete disorder within the alkyl chain would result in effective inversion symmetry. More recently, Molecular Dynamics simulations have revealed an alternative explanation, which relies on a restoration of order causing the transition.

Details of these simulations can be found elsewhere [14], but the most important finding is depicted in Fig. 3. These simulations demonstrate that at sufficiently low lipid concentration (>~110 Å^2/molecule), the monolayer ruptures, and pores appear. As a result, local domains appear (circle in Fig. 3) with high alkyl chain order – much like the compressed phase. As the signal from both these ordered domains and the water phase will be very small, this could explain the disappearance of the CH_2 signal.

3. CARS microscopy of vesicles.

Whereas SFG is well suited to study lipid molecular conformations in monolayer systems, CARS microscopy is a great tool to image the molecular state of lipids in multi-lamellar vesicles with very high spatial resolution [10,11]. Multi-lamellar vesicles are good model systems for cell membranes, and owing to their engineered multi-lamellar structure, they are well-suited for investigations of the phase behavior of lipids constituting the vesicle. This is clearly illustrated in Figure 4, which shows a typical (multiplex) CARS microscopic image of two vesicles made of different lipids (DSPC and DOPC, respectively), where the Raman spectral signature is characteristic of the phase of the lipid in the vesicle (i.e. with ordered alkyl chains in the gel-phase or with disordered chains in the liquid crystalline lipid membrane) [16]. The figure shows a multiplex CARS image with both a gel phase DSPC vesicle and a liquid crystalline-phase DOPC vesicle within the field of view. This image is obtained within a single scan of the specimen, where the discrimination between the two types of vesicles is based only on differences in the CARS spectra obtained at each specimen position. Typical spectra, with an acquisition time of 50 ms, obtained for DSPC and DOPC are shown on the right hand of the figure. The spectral data at each point were subsequently fitted to theoretical expressions for the CARS signals of a gel-phase lipid (DSPC) and a liquid crystalline-phase lipid (DOPC). The contrast in the different representations of the vesicles is based on the amplitude of the -1128 cm^{-1} skeletal mode of the gel-phase lipids and the -1087 cm^{-1} mode of the liquid crystalline-phase lipids, respectively, as derived from the fitting of the data.

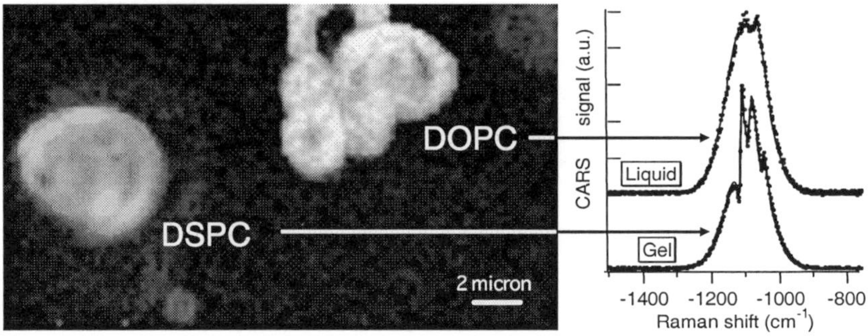

Fig. 4. Multiplex CARS image of both a DSPC and a DOPC vesicle, which have been discriminated on the basis of their CARS spectrum in the skeletal optical mode region around 1100cm^{-1}. The image is 100×60 pixels, with a 50-ms acquisition time per pixel. The scale bar corresponds to 2 μm.

4. Conclusions and outlook

We have attempted to illustrate the strength of applying non-linear optical techniques that are vibrationally specific to the study of bio-mimetic surfaces. It is evident that both SFG – with its unique surface specificity and sensitivity to molecular order – and CARS microscopy – with, a.o. its excellent spatial resolution, are promising techniques that will allow the study of increasingly complex membrane systems. Indeed, very recent developments in both areas hold promise for the future: SFG has recently been demonstrated to applicable also to non-planar surfaces [17]. This has opened new ways for investigating biologically relevant interfaces in a highly detailed fashion.

Despite its inherent insensitivity to the surface layer, CARS spectroscopy has recently been shown to also work on single lipid monolayers [18]. In this case the contrast with potential signal from the bulk can be made on the basis of the vibrational modes (C—H modes) that are selectively present in the monolayer, and not in the underlying water phase.

Acknowledgments

This work is part of the research program of the Foundation for Fundamental Research on Matter (FOM), which is financially supported by the Netherlands Organization for Scientific Research (NWO). The authors are grateful to Sylvie Roke, Aart Kleyn, Juleon Schins, Sjors Wurpel, Fred Brakenhoff, Nicoleta Nastase, Hilde Rinia, Siewert-Jan Marrink, and Alan Mark for their invaluable contributions to this work and continuous support, and Ellen Backus for carefully reading the manuscript.

References

1. M. Edidin, *Mol. Cell Biol.* **4**, 414418 (2003).
2. K. Simons and W. L. C. Vaz, *Annu. Rev. Biophys. Biomol. Struct.* **33**(1), 269-295 (2004).
3. Y. R. Shen, The principles of nonlinear optics. New York: John Wiley & Sons (1984)
4. M. Schubert and B. Wilhelmi, Nonlinear optics and quantum electronics. New York: John Wiley & Sons (1986)
5. R. W. Boyd, Nonlinear optics. Boston: Academic Press, Inc. (1992)
6. S. Mukamel, Principles of nonlinear optical spectroscopy. Oxford: Oxford University Press (1995)
7. C. Flytzanis: Theory of nonlinear optical susceptibilities. In: Quantum Electronics, vol. I: Nonlinear Optics, H. Rabin and C. L. Tang, Eds. New York: Academic Press, 1975, pp. 9-207.
8. Y. Prior, *IEEE J. of Quant. El.* **20**, 37 (1984).
9. Y.R. Shen, Nature 337 (1989) 519 and references therein.
10. A. Zumbusch, G. R. Holtom, and X. S. Xie, *Phys. Rev. Lett.* **82**, 4142 (1999).
11. M. Müller, J. Squier, C. A. de Lange, and G. J. Brakenhoff: J. *Microscopy* **197**, 150 (2000).
12. V.M. Kaganer, H. Möwald and P. Dutta, *Rev. Mod. Phys.* **71**, 779 (1999).
13. S. Roke, J. Schins, M. Muller and M. Bonn, *Phys. Rev. Lett.* **90**, 128101 (2003).
14. V. Knecht, M. Müller, M. Bonn, S.-J. Marrink, and A. E. Mark, *submitted.*
15. P.B. Miranda, V. Pflumio, H. Saijo and Y.R. Shen, *Chem. Phys. Lett.* **264**, 387 (1997); *J. Am. Chem. Soc.* **120**, 12092 (1998).
16. M. Müller and J. M. Schins, *J. Phys. Chem. B* **106**, 3715 (2002).
17. S. Roke, W.G. Roeterdink, J.E.G.J. Wijnhoven, A.V. Petukhov, A.W. Kleyn and M. Bonn, *Phys. Rev. Lett.* **91**, 258302 (2003).
18. G.W.H. Wurpel, J.M. Schins and M. Müller, *J. Phys. Chem. B* **108**, 3400 (2004).

AB INITIO STUDY OF THE GE(111):SN SURFACE

PAOLA GORI[1], OLIVIA PULCI[2], ANTONIO CRICENTI[1]

[1] *Istituto di Struttura della Materia, Consiglio Nazionale delle Ricerche, Via del Fosso del Cavaliere 100, I-00133 Roma, Italy,*

[2] *INFM, Dipartimento di Fisica dell'Università di Roma "Tor Vergata", Via della Ricerca Scientifica 1, I-00133 Roma, Italy*

We present a Density Functional Theory calculation of the geometrical and electronic structure of Sn on Ge(111) surface. The 3×3 and $\sqrt{3} \times \sqrt{3}$ reconstructions are studied in details, performing also band structure calculation and simulating STM images. Our results confirm that the 1U-2D 3x3 structure is the more favorable one, from an energetic point of view but also from a comparison with the available photoemission and STM experiments. The 'static' room temperature electronic band structure of the $\sqrt{3} \times \sqrt{3}$ instead hardly compares with the available photoemission data, thus supporting the idea of a dynamical flipping at room temperature of the Sn ad-atoms.

1. Introduction

Adsorption of metals on semiconductor surfaces and metal-semiconductor interfaces are the subject of many research efforts because of their interest from both a fundamental and an applicative point of view. In particular, attention has been devoted to the study of the adsorption of Pb or Sn on Si and Ge surfaces. The α-phase (1/3 monolayer adsorbate coverage) of Sn on Ge(111) (see Fig. 1 for the geometry of the problem) undergoes a reversible phase transition from $(\sqrt{3} \times \sqrt{3})$ R 30° to 3×3 when the temperature is lowered below 200 K [1,2], whereas no phase transition takes place for the α-phase of Sn on Si(111) down to 6 K [3].

For Sn on Ge(111), it was suggested [4,5,6] that the $\sqrt{3} \times \sqrt{3}$ symmetry phase found at room temperature is actually the result of a rapid dynamical flipping between different Sn configurations, which are defined by the static 3×3 ground state symmetry. The configuration of the three Sn ad-atoms is generally accepted to be 1U-2D (one Sn up, two Sn down) [7,8], but recently this picture has been questioned by Davila et al. [9] who suggest a 2U-1D (two up, one down) geometry. Conflicting results arise also from the interpretation of Surface X-Ray Diffraction (SXRD) data concerning the room-temperature $\sqrt{3} \times \sqrt{3}$ phase: some of them [11] support a corrugated

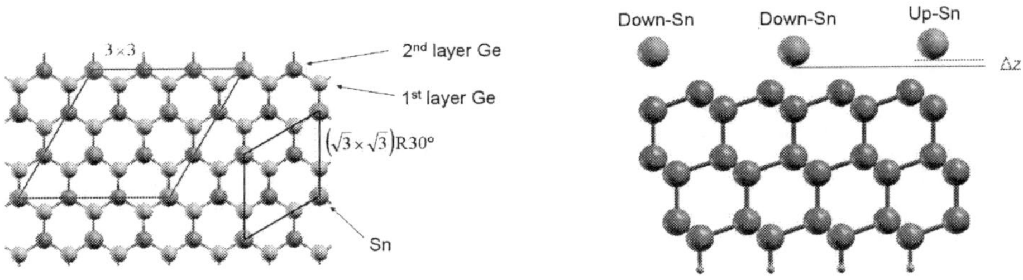

Figure 1. Top and side view of the atomic geometry of the Sn/Ge(111) surface. The two possible reconstructions are indicated in the top view, whereas the side view is shown for the surface 3×3 1U-2D.

structure, whereas other results [10] indicate a flat Sn layer.

In this paper we analyze the possible configurations by performing well converged total energy calculations with density functional theory (DFT) [12] in the local density approximation (LDA) [13]. We present also electronic band structure calculations and simulated STM images and compare our results with available experimental data.

2. Method

First-principles calculations [14] have been performed within the DFT-LDA approximation using norm-conserving pseudopotentials [15]. Three configurations for Sn ad-atoms have been analyzed: a $\sqrt{3} \times \sqrt{3}$ structure, where all the ad-atoms have the same height, and two types of 3×3 reconstructions: the 1U-2D and the 2U-1D.

All systems were modelled by a slab made of 6 Ge layers, plus one layer of H atoms saturating the bottom Ge layer and plus Sn atoms on the top of the slab. Relaxation has been performed in all cases keeping the last bottom Ge layer and the saturating H atoms fixed, until the Hellmann-Feynman forces on all other atoms were reduced below $7 \cdot 10^{-3}$ eV/a.u. .

A plane wave basis set with 12 Ry energy cutoff has been used. The Surface Brillouin Zone (SBZ) of the 3×3 phase has been sampled by a 15×15 k-points grid.

3. Results

The relaxation of the slab atomic positions leads to three different energy local minima depending on the starting point in the configuration space. The global minimum is found for the 3×3 reconstructed surface with a 1U-2D configuration. The corresponding vertical displacement between the

'up' atom and the 'down' atoms amounts to 0.37Å. This configuration shows an energy gain of 6 meV/adatom compared with the other (local) energy minimum obtained for the 2U-1D reconstruction, where the vertical rippling is of 0.22Å. The 1U-2D reconstructed surface is also energetically favored against the relaxed $\sqrt{3} \times \sqrt{3}$ surface by 7.5 meV/adatom.
For the three geometries studied, we have also performed the DFT band structure calculations. Fig. 2 presents the result for the $\sqrt{3} \times \sqrt{3}$ surface along symmetry lines of the 3×3 SBZ. The band structures calculated for the 1U-2D and 2U-1D configurations are shown in Figs. 3 and 4 respectively. In all figures, shaded regions denote bulk-projected bands. It can be observed that the calculated 1U-2D electronic structure compares reasonably well with valence band photoemission experiments [16], whereas a much less good agreement exists between the 2U-1D calculated and measured band structure: for example, at the K point one surface band is shown by experiments at an energy of 0.2 eV below Fermi level, but no occupied surface state is found at K by the theory.

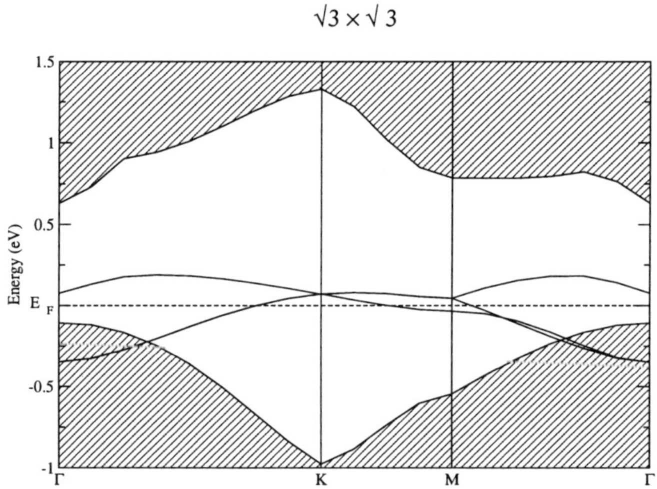

Figure 2. Electronic band structure for Sn/Ge(111) with $\sqrt{3}\times\sqrt{3}$ surface configuration, along k-points of the 3×3 SBZ

Finally, it is interesting to notice that the calculated electronic structure of the $\sqrt{3} \times \sqrt{3}$ surface shows, instead, a trend markedly different from that observed in experimental room temperature bands, e.g. at the K

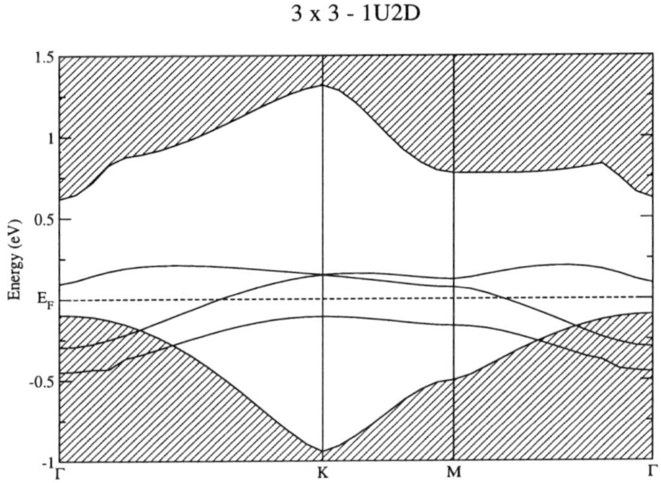

Figure 3. Electronic band structure for Sn/Ge(111) with 1U-2D 3×3 surface reconstruction

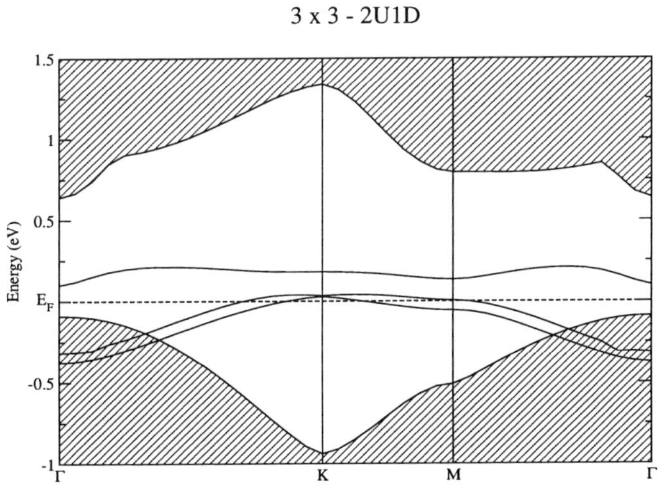

Figure 4. Electronic band structure for Sn/Ge(111) with 2U-1D 3×3 surface reconstruction.

point, where the minimum found in the experiments does not appear in the theory. This discrepancy is very important since it can be considered a point supporting the idea that the $\sqrt{3} \times \sqrt{3}$ phase has a non-static nature but results from a dynamical flipping between 3×3 configurations. This picture is also supported by the strong similarity observed between the measured valence band structures at low temperature and at room temperature [16].

If the dynamical picture is accepted, an association between the inequivalent Sn ad-atoms and the surface bands remains to be done. This is illustrated by Fig. 5 which shows, for the 1U-2D reconstructed surface, the spatial location of empty and filled states at the K point. Circles indicate positions of the atoms. From top to bottom, what is shown is the squared wavefunction at the K point of the filled surface band, the first empty band and the second empty band. It is seen that filled states are localized on the up Sn and empty states on the down Sn.

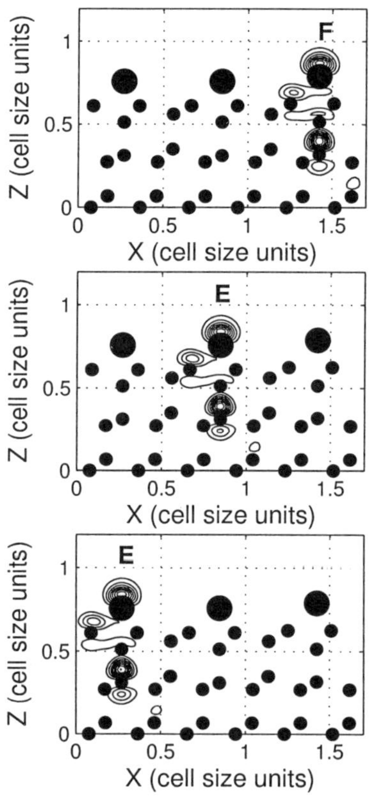

Figure 5. Spatial location of empty and filled states at the K point in the SBZ of the 1U-2D 3 × 3 surface.

A further comparison between theory and experiments can be performed by considering simulated STM images for the three considered configurations. The images have been obtained from the integrated local density of states, using the Tersoff-Hamann approach [17].

Images shown in Fig. 6 refer to a bias value of ±0.5 V applied between the sample and the STM tip and are in constant-height mode. The tip-sample distance is 2 Å both for filled and empty states. Because of the gap underestimation found within DFT already for the Germanium bulk (see Tab.1), we can safely say that bias values used for simulating STM images have to be contrasted with larger experimental values.

It can be noticed that only images obtained for the 1U-2D structure are compatible with actually observed low-temperature STM images (see, e.g., [2]), which show an hexagonal lattice for filled states and a honeycomb pattern for empty states. Filled states of simulated 2U-1D structures are, instead, complementary to what observed in experiments, whereas a static $\sqrt{3} \times \sqrt{3}$ surface would provide the same STM image with positive or negative bias, i.e. it would give a real structural information, which is not the case in practice.

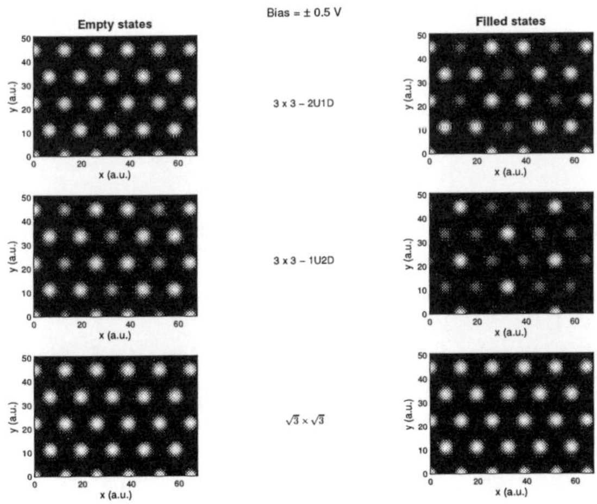

Figure 6. Calculated STM images of the three considered surface structures.

Table 1. Calculated (within DFT and within GW [18]) vs. experimental [19] energy gaps at high-symmetry points for Germanium bulk.

	Energy difference (eV)		
k-points	DFT	GW	Experimental
B5(L)-B4(Γ)	0.33	0.8	0.8
B5(Γ)-B4(Γ)	0.50	1.1	1.0
B5(L)-B4(L)	1.78	2.3	2.2
B5(X)-B4(X)	3.88	4.4	4.45

4. Conclusions

The electronic structure of 1/3 ML Sn/Ge(111) has been studied within DFT and different possible equilibrium structures have been compared. Our results concur to the idea that the transition $3 \times 3 \rightarrow \sqrt{3} \times \sqrt{3}$ is not a real phase transition but a dynamical effect of averaging on the 3x3 1U-2D configurations, which disappears when the temperature is lowered. Moreover, among the two possible 3×3 configurations, only the 1U-2D appears to be compatible with low-temperature experimental results.

Acknowledgments

This work has been supported by the INFM PAIS project "CELEX", MIUR-COFIN 2002 and by the EU's 6th Framework Programme through the Network of Excellence 'NANOQUANTA' (NMP4-CT-2004-500198). We gratefully acknowledge CINECA CPU time granted by INFM.

References

1. J.M. Carpinelli, H.H. Weitering, E.W. Plummer and R. Stumpf, Nature **381** 398 (1996).
2. M. Carpinelli, H.H. Weitering, M. Bartkowiak, R. Stumpf and E.W. Plummer, Phys. Rev. Lett. **79** 2859 (1997).
3. H. Morikawa, I. Matsuda, S. Hasegawa, Phys. Rev. B **65** 201308 (2002).
4. J. Avila, A. Mascaraque, E.G. Michel, M. C. Asensio, G. Le Lay, J. Ortega, R. Perez and F. Flores, Phys. Rev. Lett. **82** 442 (1999).
5. J. Ortega, R. Perez and F. Flores, J. Phys.: Condens. Matter **12** L21 (2000).
6. L. Jurczyszyn, J. Ortega, R. Perez and F. Flores, Surf. Sci. **482-485** 1350 (2001).
7. S. de Gironcoli, S. Scandolo, G. Ballabio, G. Santoro and E. Tosatti, Surf. Sci. **454-456** 172 (2000).
8. G. Ballabio, G. Profeta, S. de Gironcoli, S. Scandolo, G.E. Santoro and E. Tosatti, Phys. Rev. Lett. **89** 126803 (2002).
9. M.E. Davila, J. Avila, H. Ascolani, G. Le Lay, M. Gothelid, U.O. Karlsson and M. C. Asensio, Appl. Surf. Sci **234** 274 (2004).

10. O. Bunk, J. H. Zeysing, G. Falkenberg, R.L. Johnson, M. Nielsen, M.M. Nielsen and R. Feidenhans'l, Phys. Rev. Lett. **83** 2226 (1999).
11. J. Avila, A. Mascaraque, G. Le Lay, E.G. Michel, M. Gothelid, H. Ascolani, J. Alvarez, S. Ferrer and M.C. Asensio, hhtp://arXiv.org/abs/cond-mat/0104259.
12. P. Hohenberg and W. Kohn, Phys. Rev. **136**, B864 (1964).
13. W. Kohn and L. J. Sham, Phys. Rev. **140**, A1113 (1965).
14. Calculations have been performed using the PWSCF code: S. Baroni, S. de Gironcoli, A. Dal Corso, P. Giannozzi, http://www.pwscf.org/
15. N. Trouiller and J.L. Martins, Phys. Rev. B **43** 1993 (1991).
16. R. I. G. Uhrberg and T. Balasubramanian, Phys. Rev. Lett. **81** 2108 (1998).
17. J. Tersoff and D.R. Hamann, Phys. Rev. B **31** 805 (1985).
18. L. Hedin, Phys. Rev. **139** A796 (1965).
19. Landolt-Börnstein, *Physics of group IV elements and III-V compounds*, Springer-Verlag, Berlin (1982).

Lifetimes of excited states

B. Hellsing
Department of Physics, Göteborg University,
S-412 96 Göteborg, Sweden

Abstract

Many important chemical and physical phenomena are influenced by inherent dissipative processes which involve energy transfer between the electrons (electron-electron scattering) and between the electrons and the ionic motion (electron-phonon scattering). The non-adiabatic interaction between the valence electrons and the ion motion in a solid reveals the break down of the Born-Oppenheimer approximation. To pin down the influence of the electron and phonon structure on these scattering processes the two-dimensional surface states are ideal both from an experimental and theoretical point of view. Several experimental techniques presently in use are able to give information about the lifetime of an excited electron or hole in the surface state band. With help from advanced theoretical calculations it is possible to sort out the relative importance of the electron-electron and electron-phonon scattering processes responsible for the quenching of the excitation and to point out the key parameters of the electron and phonon structure.

1 Introduction

In the spirit of the Born-Oppenheimer approximation we consider that the electrons respond instantaneously to the motion of the ions in a metal. The argument is that the electrons are light in comparison and move rapidly. However, the electrons do not move infinitely fast and thus if an ion is displaced the electron system will for a short but finite time be in an excited state. Is this something we have to be worried about ? Well, it depends on what we are concerned with. To obtain information about crystal structure and in general also the electron structure - as long as we are interested in ground state properties such as electron densities and density of states - the Born-Oppenheimer picture works fine.

In surface science, adsorption, desorption and reaction often proceeds via an intermediate electronically excited state. From a theoretical point of view these dynamical surface processes points to two fundamental tasks, to calculate from *first principles* (1) Potential energy surfaces (PES) of excited states and (2) energy dissipation due to non-adiabatic coupling between nuclear motion and the electron density. Our aim is to obtain further understanding of the non-adiabatic coupling and to point out relevant calculation schemes and experiments for this purpose.

Having in mind a typical dynamical surface process, one should ask the question if the lifetime of the excited state is compatible with the time it takes for the adsorbate to leave the surface, or at least long enough that a significant fraction will. For this reason it is important to find the parameters that determine the lifetime of the excitation. The way we experimentally can get a grip of the non-adiabatic coupling between the ionic motion and the conduction electrons is by performing a well defined excitation of the system, either by a vibrational excitation or an electronic excitation and then monitoring, by some means, the decay time to the ground state. In the first case, considering for instance a molecule adsorbed on a metal surface, the observed vibrational line width, Γ, in an Electron energy loss spectroscopy (EELS) or Infra red spectroscopy (IRS) measurement give information about the vibrational lifetime, $\tau = \hbar/\Gamma$. Considering an electronic excitation Photoemission spectroscopy (PES) and lately also Scanning tunneling spectroscopy (STM) are able to give information about the lifetime of an excited one-electron state.

In this paper we will discuss the decay of a hole state in a surface state band. The reason why there has been such a focus on surface states in this context is rather obvious as from an experimental and theoretical point of view these states are excellent for benchmark studies. The influence of the solid on the photo electron in PES is

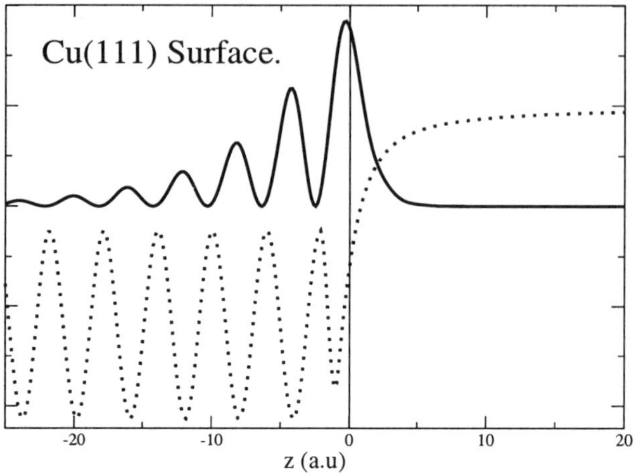

Figure 1: Surface state wave function, absolute magnitude squared (solid line) and the model potential (dotted line) for the Cu(111) surface.

minimized and the surface states, two dimensional in character, are for many systems well isolated from bulk bands in a large fraction of the surface Brillouin zone. In Fig. 1 the surface state wave function for Cu(111) is shown. It is seen that the wave function decays exponentially in the direction toward the bulk within about 4 atomic layers. Today, several ways to experimentally investigate lifetimes of surface localized electron states have been presented, Angular resolved photoemission electron spectroscopy (ARPES) [1, 2, 3, 4, 5, 6, 7], Two-photon photoemission (2PPE) techniques [8, 9], Time resolved two-photon photoemission technique for direct determination of excitation lifetimes [10, 11, 12, 13, 14, 15, 16, 17, 18, 19, 20, 21] and different methods based on Scanning tunneling spectroscopy [22, 23, 24, 25, 26, 27, 28].

2 Inherent scattering processes

The decay channel of the hole in the surface state band via excitations of secondary electrons, the so called electron-electron (e-e) interaction has been investigated theoretically in detail for many systems [29, 30, 31]. In Fig. 2 the e-e scattering is shown schematically. The figure illustrate an event when a bulk electron, due to the screened interaction V_{ee} with an other electron, is scattered into the hole of the surface state band and simultaneously an electron is scattered from the occupied part of the surface state band into the unoccupied part of a bulk band in order to conserve momentum an energy. The non-adiabatic ion-electron interaction, or as more often phrased, the electron-phonon coupling e-p, is illustrated in Fig. 3. In this case, what drives the scattering of an electron into the hole is the field, generated by the vibrating ions giving rise to a change in the screened ion-electron interaction, δV_{ie}. In this case we have to sum up all events including phonon absorption and emission which fulfill energy and momentum conservation.

3 Qualitative discussion of non-adiabatic ion-electron coupling

Before giving the detailed expressions for the phonon induced lifetime broadening Γ_{ep} we can discuss qualitatively the expected binding energy and temperature dependence.

3.1 Binding energy dependence

For the e-p scattering the energy window of scattering events is twice the maximum phonon energy, ω_{max}, typically 40-60 meV (indicated in Fig. 3). Thus we should expect that the phonon induced lifetime broadening Γ_{ep} versus hole binding energy - energy below Fermi level - essentially will saturate for binding energies exceeding the ω_{max}.

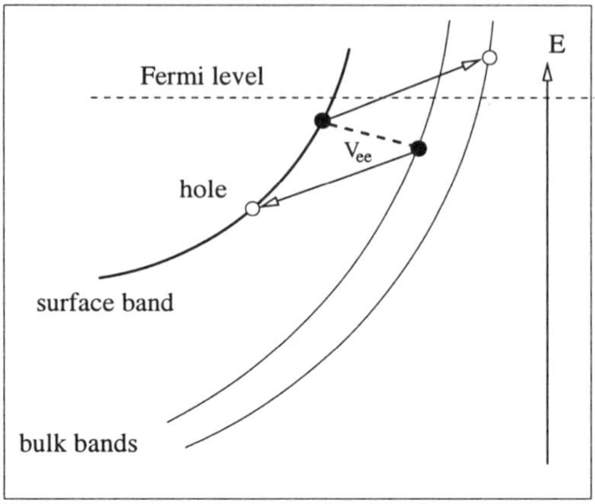

Figure 2: Schematic illustration of the energy and parallel momentum conservation in an electron-electron scattering event which leads to filling of the photo hole in the surface state band. V_{ee} denotes the screened electron-electron interaction.

3.2 Temperature dependence

For a qualitative discussion we can start off with a simple one dimensional model. Consider a single ion in an electron gas. If the ion is displaced a typical vibrational amplitude from the equilibrium position ($R = 0$) we can expand the screened ion-electron interaction potential V_{ie} to first order and the Hamiltonian is written

$$H = H_o + \delta V_{ie} = H_o + \frac{\partial V_{ie}(x)}{\partial R} R \, , \tag{1}$$

Where x is the electron coordinate and the derivative of the ion-electron potential is evaluated at the equilibrium position ($R = 0$). The solutions to H_o are the adiabatic states, $|I, n >= \Psi_I(R)\phi_n(R, x)$, where $\Psi_I(R)$ is the wave function of the ion and $\phi_n(R, x)$ the electron wave function which depends parametrically on the ion position.

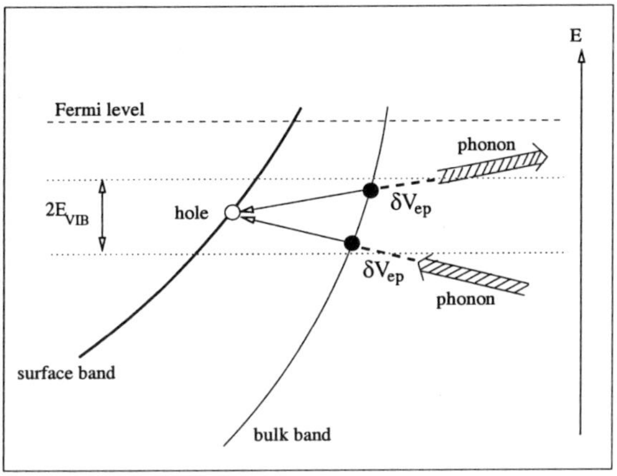

Figure 3: Schematic illustration of the energy conservation in electron-phonon scattering events, phonon absorption and emission, which leads to filling of the photo hole in the surface state band. δV_{ie} denotes the change of the screened ion-electron interaction V_{ie} due to the vibrational motion.

Consider now the initial state where an electron is photo emitted, leaving behind a hole. Applying the "Golden-rule", first order time dependent perturbation theory, we can calculated the electron scattering rate into the hole state $n = n_i$, assuming the ion is initially in the vibrational state I.

$$\Gamma_{I,n_i} = 2\pi \sum_{J,n} | < J, n | \frac{\partial V_{ie}(x)}{\partial R} R | I, n_i > |^2 \delta(\hbar\omega_{IJ} - \hbar\omega_{nn_i}) =$$

$$\sum_J | < I|R|J > |^2 \sum_n | < n | \frac{\partial V_{ie}(x)}{\partial R} | n_i > |^2 \delta(\hbar\omega_{IJ} - \hbar\omega_{nn_i}) \ , \quad (2)$$

where $\hbar\omega_{IJ} = E_I - E_J$ and $\hbar\omega_{nm} = \epsilon_n - \epsilon_m$. We assume the vibrating ion to be a harmonic oscillator with a single frequency Ω_o. If now consider for example processes involving emission of one single vibrational quanta $\hbar\Omega_o$, we have $J = I + 1$. Having the oscillating ion in thermal equilibrium with a temperature T we take the expectation value of the vibrational quantum number I, $< I >= n_B(T)$ where $n_B(T) = [exp(\hbar\Omega_o/k_BT) - 1]^{-1}$. This will give us the essential part of the temperature dependence. Later on we we will also include the temperature dependent Fermi factors of the electrons, but unless temperature are very high or we are concerned with states very close to the Fermi level these are of no importance. So, assuming that the binding energy of the hole state exceeds $\hbar\Omega_o$ and summing up electron scattering events which involves emission of a vibration quanta we have

$$\Gamma_{n_i} = 2\pi\frac{\hbar}{2M\Omega_o}[1 + n_B(T)]\sum_n | < n|\frac{\partial V_{ie}(x)}{\partial R}|n_i > |^2\delta(\hbar\Omega_o - \omega_{nn_i}) \ , \quad (3)$$

where M is the ion mass. In the case of absorption of a vibrational quanta the temperature factor $1+n_B(T)$ is replaced by $n_B(T)$. Thus we can deduce the qualitative temperature dependence. In the limit $k_BT << \hbar\Omega_o$ we have $\Gamma_{n_i} \sim 1$ and for high temperatures $k_BT >> \hbar\Omega_o$ we obtain $\Gamma_{n_i} \sim T$. Thus for low temperatures Γ_{n_i} takes a finite value while at high temperatures it becomes linear with temperature. This qualitative behavior can been seen in Fig. 5.

Now, we proceed and consider a solid terminated by a surface. We will apply the slab model which means we consider a finite number of atomic layers and periodic boundary conditions in the lateral directions. From the point of view of vibrational properties we now have coupled vibrating ions which yields collective vibrational modes, phonons. Applying a simple single force constant model, we consider springs attached between neighboring ions and then we calculate the eigen vectors (phonon polarization vectors) and eigen values (phonon dispersion relation) of the dynamical matrix. The phonons will be characterized by a mode index ν and a momentum, $\vec{q}$ parallel to the atomic layers of the slab. In a metal slab model we have one-electron wave functions and energies

$$\psi_{n,\vec{k}_{\|}}(z, \vec{x}) = \frac{1}{\sqrt{A}}\ \phi_n(z)\ e^{i\vec{k}_{\|}\cdot\vec{x}} \quad , \quad \epsilon_n(\vec{k}_{\|}) = \epsilon_n^0 + \hbar^2 k_{\|}^2/2m_n, \quad (4)$$

where n is the band index, $\vec{k}_{\|}$ the momentum parallel to the surface and A the surface area. In the following we will suppress the $\|$ index. The z coordinate is along the surface normal and $\vec{x}$ in the surface plane.

4 Phonon induced lifetime broadening

Within our slab model we now outline the equations for calculating the phonon induced lifetime broadening of a hole in a surface state band. Summing up the contributions from phonon emission, corresponding to Eq. (3) in the case of an isolated vibrating ion, and phonon absorption and also taking into account the temperature dependence of the occupancy of the electron states, f, we have

$$\begin{aligned} \Gamma_{ep}(\omega, \vec{k}_i) &= 2\pi \sum_{\nu,\vec{q},f} |g^{\nu}_{i,f}(\vec{q})|^2 \times \\ \{ \; [1 + n_B(\hbar\omega_\nu) - f(\epsilon_{f,\vec{k}_i-\vec{q}})] \; & \delta(\hbar\omega - \epsilon_{f,\vec{k}_i-\vec{q}} - \hbar\omega_\nu) \; + \\ [n_B(\hbar\omega_\nu) + f(\epsilon_{f,\vec{k}_i-\vec{q}})] \; & \delta(\hbar\omega - \epsilon_{f,\vec{k}_i-\vec{q}} + \hbar\omega_\nu) \;] \; \} \; . \end{aligned} \tag{5}$$

Eq. (5) is obtained from a Master type of equation where the time dependent filling of the initially empty hole state is determined by the scattering rate into the hole minus the scattering rate out of the hole [38]. The so called electron-phonon coupling function squared is given by

$$|g^{\nu}_{i,f}(\vec{q})|^2 = \frac{\hbar}{2MN\omega_\nu(\vec{q})} \times |\langle \, f \mid \sum_{\mu} \vec{\epsilon}_{\vec{q}\nu}(\vec{R}_\mu) \cdot \vec{\nabla}_{\vec{R}_\mu} \tilde{V}^{\mu}_{q} \mid i \, \rangle|^2 \; . \tag{6}$$

Comparing Eqs. (5) -(6) and Eq. (3) we note the similar structure. We consider the static screening of the electron-ion potential, since the phonon frequencies are in general small in comparison with the energies of the scattered electrons. The coupling function in Eq. (6) is the result of the standard first order expansion of the screened electron-ion potential, $\tilde{V}^{\mu}_{q}$, with respect to the vibrational coordinate $\vec{R}_\mu$. N is the number of ions in each layer, M is the ion mass, μ is the layer index and $\vec{\epsilon}_{\vec{q}\nu}(\vec{R}_\mu)$ are the phonon polarization vectors.

To relate to a conceptually more simple picture of the phonons we introduce the Eliashberg spectral function $\alpha^2F(\omega)$ [32] which is the phonon density of states weighted by the *e-p* coupling function g

$$\alpha^2 F_{\vec{k}_i}(\omega) \ = \sum_{\nu,\vec{q},f} |g^{\nu}_{i,f}(\vec{q})|^2 \ \delta(\hbar\omega - \hbar\omega_\nu(\vec{q})) \ \delta(\epsilon_f - \epsilon_i) \ , \tag{7}$$

where the last delta function indicates that we consider the quasi-elastic approximation, neglecting the change of the energy of the scattered electron due to absorption or emission of a phonon. Thus we can write Γ_{ep} as an integral over phonon energies. If we consider an initial hole state $(\omega, \vec{k}_i)$ and take into account phonon absorption and emission processes we obtain

$$\begin{aligned} \Gamma_{ep}(\omega, \vec{k}_i) \ &= 2\ \pi \int_0^{\hbar\omega_{max}} \ \alpha^2 F_{\vec{k}_i}(\epsilon) \times \\ [1 + 2n_B(\epsilon) &+ f(\hbar\omega + \epsilon) - f(\hbar\omega - \epsilon)] \ d\epsilon \ , \end{aligned} \tag{8}$$

where ω_{max} is the maximum phonon frequency. We then obtain the $T = 0$ result ($\Rightarrow n_B = 0$) for Γ_{ep} as a function of hole binding energy $|\omega|$.

$$\Gamma_{ep}(\omega, \vec{k}_i) \ = 2\pi \int_0^{|\omega|} \alpha^2 F_{\vec{k}_i}(\epsilon) d\epsilon \ . \tag{9}$$

The mass enhancement parameter λ is in terms of the Eliashberg function just its first reciprocal moment[33]

$$\lambda(\vec{k}_i) = 2 \int_0^{\omega_m} \frac{\alpha^2 F_{\vec{k}_i}(\omega)}{\omega} \ d\omega \ . \tag{10}$$

If the high T-limit ($k_BT >> \hbar\omega_{max}$) of Eq. (8) is considered, Grimvall [34] has pointed out a very useful result which enables an experimental determination of the mass enhancement parameter

$$\Gamma_{ep}(\omega, \vec{k}_i) \ = 2\pi \ \lambda(\vec{k}_i) \ k_BT \ . \tag{11}$$

We thus conclude that the Eliashberg function α^2F is a basic function to calculate. Given this function most of the interesting quantities can be calculated, such as the temperature and also binding energy dependence of the lifetime broadening and the mass enhancement parameter. However, this is no simple task, as all the physics connected to the *e-p* interaction is buried in $\alpha^2 F$, the phonon dispersion relation, phonon polarization vectors, one-electron wave functions and last but not least, the gradient of the screened electron-ion potential - the deformation potential.

5 Calculations

We here present results from a recent calculation of the lifetime broadening of the intrinsic surface states of the noble metal surfaces Cu(111) and Ag(111) compared to High resolution ARPES data [36]. In Fig. 2 the schematic band structure is presented.

Considering the *e-p* interaction the aim was to take into account in some details the phonons, both bulk and surface modes. To achieve the phonon dispersion and the phonon polarization vectors we performed a slab calculation [36]. We adopted the Ashcroft pseudo potentials as bare electron-ion potentials, parametrized according to Ashcroft and Langreth [37] . We have investigated the screening of the bare potentials by applying the dielectric function according to both Thomas-Fermi and RPA (constructed by the eigen wave functions and energy eigen values from a 31-layer slab calculation). The two different types of screening gave a difference of about 1 % for the mass enhancement factor (λ) and the phonon induced lifetime broadening (Γ_{ep}) due to compensating effects [35].

Calculating Γ_{ep} we have to take into account intra band and inter band scattering of electrons and also possibly Umklapp processes. For the surface states of the studied noble metals the Umklapp processes can be neglected as the Fermi momentum of the surface states are small ($< 0.12\ a.u.$) in comparison with half the minimum reciprocal vector ($|\vec{G}|/2 < 0.75\ a.u.$) [35]. Furthermore the intra band scattering has been shown by us to be of minor importance [38].

We now turn to the results of the calculations concerning the hole binding energy and temperature dependence of the studied surface states. In Fig. 4 we present the calculated hole binding energy dependence of Γ_{ep}, at $T = 30K$, for the surface states of Cu(111) and

Ag(111). The calculated structure of Γ_{ep} in the small binding energy region reflects the structure of the Eliashberg function which in turn depends on the phonon density of states of the system. The high resolution ARPES data show some of these structures, indicating the possibility to experimentally obtain the Eliashberg function at low temperatures. We also note, as mentioned in the introduction that the saturation of Γ_{ep} at $\omega = \omega_{max}$ ($\approx$ 30 meV for Cu(111) and 20 meV for Ag(111)) is also seen in the experimental data.

Adding the contribution from the *e-e* interaction, values close to the experiment are obtained [36]. We note from Fig. 4, that the contribution from the Rayleigh surface mode gives an important contribution in particular for very small binding energies.

The main signature of the *e-p* contribution to the lifetime broadening is the temperature dependence. In section 3.2 we argued that if the binding energy of the hole state exceeds the vibrational energies Γ_{ep} takes a finite value while at high temperatures it becomes linear with temperature. This qualitative behavior can been seen in our calculated temperature dependence of Γ_{ep} presented in Fig. 5 considering a hole state in the $\bar{\Gamma}$ point for Cu(111) and Ag(111) . The calculated full lifetime broadenings for both Cu(111) and in particular for Ag(111), are in excellent agreement with the experimental data [36].

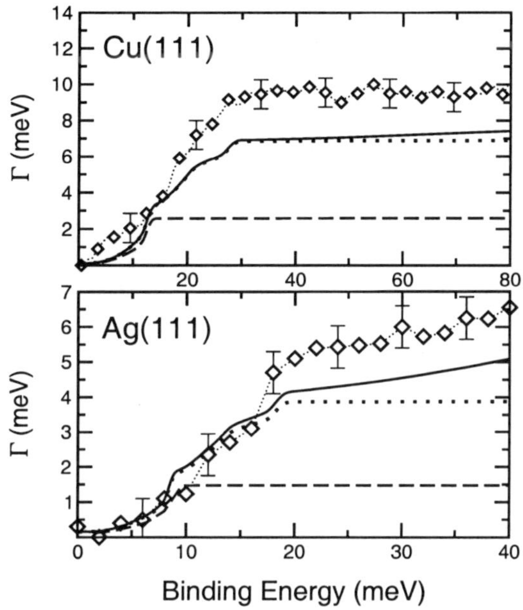

Figure 4: Lifetime broadening of the Cu(111) and Ag(111) surface hole state as a function of binding energy, $\Gamma_{ee} + \Gamma_{ep}$ (solid line), Γ_{ep} (dotted line) and the Rayleigh mode contribution to Γ_{ep} (dashed line).

6 Concluding Remarks

In this paper we give a presentation of the electron-phonon coupling starting off by describing the coupling between a single vibrating ion and the surrounding electron gas. We are then focused on evaluating the phonon induced lifetime broadening of an electronic surface band state. We have demonstrated that it is possible to reasonably well understand experimental data concerning the *e-p* induced lifetime broadening of surface states. Of major importance is to take into account bulk and surface electron and phonon states.

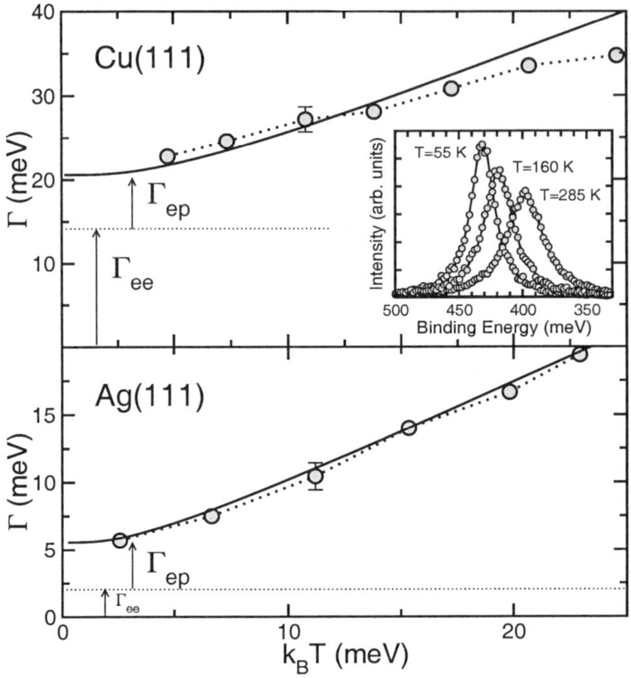

Figure 5: Lifetime broadening of the Cu(111) and Ag(111) surface hole states as a function of temperature (solid line), Γ_{ee} (dotted line).

There are presently many interesting theoretical investigations of the *e-p* interaction on metal surfaces to be done. For example to explain the observed seemingly strong *e-p* coupling for metallic multi-overlayer structures.

References

[1] S. Å. Lindgren and L. Walldén, Phys. Rev. Lett. **59**, 3003 (1987).

[2] A. Carlsson, D. Claesson, S. Å. Lindgren, and L. Walldén, Phys. Rev. Lett. **77**, 346 (1996).

[3] F. Theilmann, R. Matzdorf, G. Meister, and A.Goldmann, Phys. Rev. B **56**, 3632 (1997).

[4] R. Matzdorf, Surf. Sci. Rept. **30**, 153 (1998).

[5] J. Paggel, T. Miller, and T. Chiang, Phys. Rev. Lett. **83**, 1415 (1999).

[6] T. Valla, A. V. Fedorov, P. D. Johnson, and S. L. Hulbert, Phys. Rev. Lett. **83**, 2085 (1999).

[7] F. Reinert, G. Nicolay, S. Schmidt, D. Ehm, and S. Hüfner, Phys. Rev. B **63**, 115415 (2001).

[8] N. Fischer, S. Schuppler, R. Fischer, Th. Fauster, and W. Steinmann, Phys. Rev. **43**, 14722 (1991).

[9] N. Fischer, S. Schuppler, R. Fischer, Th. Fauster, and W. Steinmann, Phys. Rev. **47**, 4705 (1993).

[10] U. Höfer, I. L. Shumay, Ch. Reuss, U. Thomann, W. Wallauer, and Th. Fauster, Science, **277**, 1480 (1997).

[11] C. B. Harris, N. H. Ge, R. L. Lingle Jr., J. D. McNeill, and C. M. Wong, Ann. Rev. Phys. Chem. **48**, 711 (1997).

[12] M. Wolf, E. Knoesel, and T. Hertel, Phys. Rev. B **54**, R5295 (1996).

[13] H. Petek and S.Ogawa, Prog. Surf. Sci. **56**, 239 (1998).

[14] E. Knoesel, A. Hotzel, and M. Wolf, J. Electron Spectrosc. Relat. Phenom. **88-91**, 577 (1998).

[15] M. Bauer, S. Pawlik, and M. Aeschlimann, Phys. Rev. B **60**, 5016 (1999).

[16] S. Ogawa, H. Nagano, and H. Petek, Phys. Rev. Lett. **82**, 1931 (1999).

[17] A. Schäfer, I. L. Shumay, M. Wiets, M. Weinelt, Th. Fauster, E. V. Chulkov, V. M. Silkin, and P. M. Echenique, Phys. Rev. B **61**, 13159 (2000).

[18] S. Link, H. A. Dürr, G. Bihlmayer, S. Blügel, W. Eberhardt, E. V. Chulkov, V. M. Silkin, and P. M. Echenique, Phys. Rev. B **63**, 115420 (2001).

[19] X. J. Shen, H. Kwak, A. M. Radojevic, S. Smadici, D. Mocuta, and R. M. Osgood Jr., Chem. Phys. Lett. **351**, 1 (2002).

[20] W. Berthold, U. Höfer, P. Feulner, E. V. Chulkov, V. M. Silkin, and P. M. Echenique, Phys. Rev. Lett. **88**, 056805 (2002).

[21] M. Roth, M. Pickel, W. Jinxiong, M. Weinelt, and Th. Fauster, Phys. Rev. Lett. **88**, 096802 (2002).

[22] J. Li, W.-D. Schneider, R. Berndt, O. R. Bryant, and S. Crampin, Phys. Rev. Lett. **81**, 4464 (1998).

[23] L. Bürgi, O. Jeandupeux, H. Brune, and K. Kern, Phys. Rev. Lett. **82**, 4516 (1999).

[24] J. Kliewer, R. Berndt, E. V. Chulkov, V. M. Silkin, P. M. Echenique, and S. Crampin, Science **288**, 1399 (2000).

[25] J. Kliewer, R. Berndt, and S. Crampin, New J. Phys. **3**, 22.1 (2001).

[26] H. Hövel, B. Grimm, and B. Reihl, Surf. Sci. **477**, 43 (2001).

[27] K.-F. Braun and K.-H. Rieder, Phys. Rev. Lett. **88**, 096801 (2002).

[28] A. Bauer, A. Mühlig, D. Wegner, and G. Kaindl, Phys. Rev. B **65**, 075421 (2002).

[29] P. Echenique, J. Pitarke, E. Chulkov, and A. Rubio, Chem. Phys. **251**, 1 (2000).

[30] E. V. Chulkov, V. M. Silkin, and M. Machado, Surface Science **482-485**, 693 (2001).

[31] A. G. Borisov, J. P. Gauyacq, A. K. Kazansky, E. V. Chulkov, V. M. Silkin, and P. M. Echenique, Phys. Rev. Lett. **86**, 488 (2001).

[32] W. McMillan, Phys. Rev. **167**, 331 (1968).

[33] G. Mahan, in *Many-Prticle Physics, Physics of Solids and Liquids*, edited by J. Devrees *et al.* (Plenum Press, New York, 1990), p. 588.

[34] G. Grimvall, in *The Electron-Phonon Interaction in Metals, Selected Topics in Solid State Physics*, edited by E. Wohlfarth (North-Holland, New York, 1981).

[35] B. Hellsing, A. Eiguren, E. V. Chulkov, J. Phys.: Condens. Matter **14**, 5959 (2002).

[36] A. Eiguren, B. Hellsing, F. Reinert, G. Nicolay, E. V. Chulkov, V. M. Silkin, S. Hüfner, and P. M. Echenique, Phys. Rev. Lett. **88**, 066805 (2002).

[37] N. Ashcroft and D. Langreth, Phys. Rev. **159**, 500 (1966).

[38] A. Eiguren B. Hellsing, E. V. Chulkov and P. M. Echenique, Phys. Rev. **67**, 235423 (2003).

SOLITON DYNAMICS IN NON-COMMENSURATE SURFACE STRUCTURES

ALEXANDER S. KOVALEV

B.Verkin Institute for Low Temperature Physics and Engineering, 47 Lenin Avenue, Kharkov, 61103, Ukraine

IGOR V. GERASIMCHUK

Departamento de Propiedades Ópticas, Magnéticas y de Transporte, Instituto de Ciencia de Materiales de Madrid (ICMM), CSIC, Cantoblanco, 28049, Madrid, Spain

We describe analytically the nonlinear dynamics of the non-commensurate surface layer ("self-modulated" system) with a spatially periodical structure. In the framework of the Frenkel-Kontorova model the nonlinear excitations of the periodic soliton lattice, such as moving additional kinks, are investigated.

Investigation of the nonlinear dynamics of real physical systems, taking into account their discreteness, internal microstructure and spatial inhomogeneity, have always been the focus of attention in the theory of nonlinear waves and solitons, particularly periodic structures with physical parameters modulated in space ("modulated" systems), such as, for example, layered crystals. Spatial periodicity leads to a band-gap structure of the spectrum of linear waves and to the existence of the so-called "gap solitons" when the nonlinearity of the medium is taken into account [1-3]. In this paper, we investigate the systems with spatially homogeneous material parameters but spatially periodical ground state. The surface atomic layer in a non-commensurate state (see, for instance, [4-6]) is an important example of similar "self-modulated" structure. In this case, the spectrum of linear excitations has a gap structure but solitons with frequencies within a gap differ from those in the modulated media. Wc investigate analytically one-parametric topological solitons ("kinks") [7] in the gap of the spectrum of non-commensurate surface structure using the Darboux transform (see, for instance, [8]).

Let us consider a non-commensurate structure of the surface layer of atoms. We take into account the interaction between surface atoms in the harmonic approximation and assume that, in the absence of substrate, the equilibrium distance between these atoms is equal to b and differs from the interatomic distance a in a bulk. The influence of a substrate on surface atoms can be simulated by a periodical potential relief with period a. For simplicity, we approximate this relief by a trigonometric function and assume the substrate to be absolutely hard. Then the potential energy of the system is given by

$$U = \sum_n U_0 \left[1 - \cos\left(2\pi y_n / a\right)\right] + \sum_n \beta\left(y_n - y_{n-1} - b\right)^2 / 2 , \qquad (1)$$

where y_n is the position of n th atom with respect to the surface layer and β is the elastic constant in the layer. The dynamical equations for the atomic displacements $v_n = y_n - an$ in this Frenkel-Kontorova model [9] have the following form:

$$m v_{n\tau\tau} + \left(2\pi U_0 / a\right) \sin\left(2\pi v_n / a\right) + \beta\left(2 v_n - v_{n+1} - v_{n-1}\right) = 0 , \qquad (2)$$

where m is the mass of atom.

In the long-wave approximation for dimensionless variables $u = 2\pi v / a$, $x = n \cdot 2\pi\sqrt{U_0/\beta}\Big/a$, and $t = \tau \cdot 2\pi\sqrt{U_0/m}\Big/a$, we come to the well-known sine-Gordon equation (SGE) [7]:

$$u_{tt} - u_{xx} + \sin u = 0 . \qquad (3)$$

In the same approximation the total energy of the system (1) takes the form

$$U = E_0 \int dx \left[u_t^2 / 2 + u_x^2 / 2 + (1 - \cos u) + \xi u_x\right], \qquad (4)$$

where $E_0 = a\sqrt{\beta U_0}\Big/(2\pi)$, and the non-commensurability of the surface layer and substrate is characterized by the dimensionless parameter $\xi = \sqrt{\beta / U_0}\,(a - b)$. The last term ξu_x in (4) has a divergent form and does not change the form of Eq. (3) but changes the potential energy of the system and can change its ground state. In the case $b = a$, the ground state corresponds to the trivial solution of Eq. (3), $u \equiv 0$ with the energy $E = 0$. Under the condition $b \neq a$, the problem becomes more complicated.

Let us consider the case $b > a$ ($\xi < 0$) where Eq. (3) allows additional nontrivial static solutions in the following form:

$$u_0 = \pi + 2\,am(z, k) , \qquad (5)$$

where $am(z,k)$ is the elliptic amplitude with modulus k , and $z = x/k$. The solution (5) describes the "extended" system of a periodical chain of 2π -kinks ("one-dimensional dislocations" in a surface layer or fluxon lattice in a long Josephson junction) separated by the distance $L = 2kK(k)$, where $K(k)$ is the first type full elliptic integral. The width of kink expressed in terms of the initial dimensional variables is equal to $\Lambda = a^2\sqrt{\beta/U_0}\Big/(2\pi)$.

The energy density of such a periodical structure (per period) $\varepsilon = U/L$ depends on the parameter of non-commensurability ξ . For small values of this parameter, the ground state of the system is homogeneous and the periodical solution (5) can exist only under pressure conditions applied to the chain at the infinity. But when the parameter ξ exceeds a critical value $\xi_c = -4/\pi$, where $b_c = a + (4/\pi)\sqrt{U_0/\beta}$, the periodical state (5) with the modulus of elliptic function, derived from the equation $E(k)/k = \xi/\xi_c$, corresponds to the minimum of energy ($E(k)$ is the second type full elliptic integral).

Linear excitations on the background of the non-commensurate structure (5) represent the high-frequency phonon mode in the layer (upper band) and the low-frequency Swihart mode of oscillations of a kink lattice (lower band). Let us consider nonlinear excitations on this background. The elementary nonlinear excitation corresponds to an additional kink (surface dislocation) which propagates through the kink lattice (5). To obtain the exact solution for this excitation, we use the Darboux transform [8] which allows us to find more complicated solutions by using the so-called "dressing" procedure for the initial solution (5). It is simple to use the Darboux transform in the case where the initial solution depends only on one variable as in our case with $u_0 = u_0(x)$, and does not depend on time. To compact this transform, it is convenient to change over from the initial field variable $u(x,t)$ to the new variables V and W connected with the initial field u by the relations

$$V = i\left(u_x + u_t\right), \qquad W = \exp(iu), \tag{6}$$

for which Eq. (3) reads

$$V_x - V_t = (W - 1/W)/2, \qquad W_x + W_t = VW. \tag{7}$$

The associated linear problem for two complex functions $\Psi_1(x,t)$ and $\Psi_2(x,t)$ corresponds to the system (7). For the column function $\Psi = \{\Psi_1, \Psi_2\}$ and an arbitrary solution $u(x,t)$ (or $V(x,t)$ and $W(x,t)$) we have [8]:

$$4\Psi_x = \begin{vmatrix} V & \lambda + W/\lambda \\ \lambda + 1/(W\lambda) & -V \end{vmatrix} \cdot \Psi, \tag{8}$$

$$4\Psi_t = \begin{vmatrix} V & \lambda - W/\lambda \\ \lambda - 1/(W\lambda) & -V \end{vmatrix} \cdot \Psi, \tag{9}$$

where λ is the parameter of the Darboux transform. The initial Eq. (3) is the condition of consistency of the system (8,9). The solution of this system with the given initial ("seed") solution $u_0(x,t)$ (given functions $V_0(x,t)$ and $W_0(x,t)$) and an arbitrary value of λ allows us to build up the new solution $u(x,t)$ (or $V(x,t)$ and $W(x,t)$). Naturally, the parameter λ must be chosen in such a way that the real solution $u(x,t)$ can be obtained. The final relation between these two solutions reads

$$u(x,t) = u_0(x,t) - 2i\ \ln[\Psi_2(u_0,\lambda) / \Psi_1(u_0,\lambda)]. \tag{10}$$

The main task is to solve the system of linear equations with variable coefficients (8,9). In our case the problem is simplified as the initial solution (5) depends only on x, and Eqs. (9) become ordinary differential equations with constant coefficients. For the ground state (5), the functions V_0 and W_0 have the form

$$V_0 = (2i/k)\, dn(z,k)\,, \qquad W_0 = [sn(z,k) - i\, cn(z,k)]^2\,, \tag{11}$$

and the system of linear Eqs. (9) can be easily solved. We can use any arbitrary real λ to obtain a new real solution. This parameter characterizes the average velocity of the moving additional kink. The solution of Eqs. (9) has the following form:

$$\Psi_1 = a(x)\exp(\mu t) + b(x)\exp(-\mu t)\,, \Psi_2 = a(x)A^{(-)}\exp(\mu t) - b(x)A^{(+)}\exp(-\mu t)\,, \tag{12}$$

where $A^{(\pm)} = \lambda(4\mu \pm V_0)/(\lambda^2 - W_0)$ and $\mu = \pm\sqrt{(\lambda + 1/\lambda)^2 - 4/k^2}\big/4$. Positive μ corresponds to a $(0,2\pi)$ additional kink, negative one to a $(2\pi,0)$ kink.

If we substitute the solution (12) into Eqs. (8) and put the coefficients before $\exp(\pm\mu t)$ equal to zero, we find $a(x)$ and $b(x)$ and, consequently, can transform the expression for the ratio Ψ_2/Ψ_1 into the following form:

$$\Psi_2/\Psi_1 = \exp(i\rho)\,[\exp(i\varphi) - i\exp(\vartheta)] / [1 + i\exp(i\varphi + \vartheta)], \tag{13}$$

where $\tan(\rho) = 2\, sn(z,k)\, cn(z,k) / [2\, sn^2(z,k) - \lambda^2 - 1]$, $\vartheta = 2\mu[t + f(x)]$, and $\tan(\varphi) = dn(z,k)/(2k\mu)$ with

$$f(x) = (k^2/4)(\lambda^2 - 1/\lambda^2)\int_0^x dx\, [dn^2(z,k) + (2k\mu)^2]^{-1}. \tag{14}$$

The function $(\varphi \pm \rho)$ can be rewritten as $(\varphi \pm \rho) = \mp am(z,k) \pm am(z \pm \Delta/k, k)$, where the phase shift Δ of the solution depends on the parameters k and λ in the following implicit form: $k\, sn(\Delta/k,k) = 2\lambda/(\lambda^2 + 1)$.

In all above formulas we considered $\lambda_0 < \lambda < \infty$, where $\lambda_0 = (1 + k')/k$, $k' = (1 - k^2)^{1/2}$. This corresponds to a positive value of the function f, i.e. to the kink motion in the negative direction. The domain $0 < \lambda < \lambda_0$ corresponds to the opposite direction of kink motion. The function $f(x)$ in (14) can be expressed as $f(x)=x/\upsilon+\chi(x)$, where the average value of the periodical function $\chi(x)$ is equal to zero. The linear growing component of $f(x)$ determines the average velocity of a kink propagating through the non-commensurate structure:

$$< \upsilon > = 4K(k)\left\{k^2\left(\lambda^2-1/\lambda^2\right)\cdot\int_0^K dz\left[dn^2(z,k)+(2k\mu)^2\right]^{-1}\right\}^{-1} . \qquad (15)$$

Consequently, the phase ϑ in Eq. (13), $\vartheta=2\mu(x+\upsilon t)/\upsilon+2\mu\chi(x)$, describes the kink motion in the negative direction with the average velocity υ. Such a motion is accompanied by periodical oscillations at the moments when the kink propagates through each kink of the lattice. After substitution of (13) into the formula (10) we obtain the final solution describing the motion of an additional kink:

$$u(x,t)=u_0(x)-2i\ln\left\{\frac{\exp(i\kappa_+)-i\exp(\vartheta+i\kappa_-)}{\exp(-i\kappa_+)+i\exp(\vartheta-i\kappa_-)}\right\}, \qquad (16)$$

where $\kappa_\pm=(\varphi\pm\rho)/2$.

In spite of the complexity of the obtained solution, it allows a simple physical interpretation. The additional kink propagates through the non-commensurate surface structure, and this propagation is accompanied by the total deformation with the phase shift 2Δ. In the limit $\lambda\to\infty$ the kink velocity tends to its maximum value ($\upsilon\to 1$) and the phase shift tends to zero ($2\Delta\to 0$): the singular additional kink moves through the undeformed periodical structure. In the opposite limit $\lambda\to\lambda_0$ the velocity of a kink tends to its minimal value $s_0=k'K(k)/E(k)$ which coincides with Swihard velocity, the width of the kink goes to infinity, and the phase shift tends to L: the perfect non-commensurate structure rehabilitates itself. The solution (16) develops an evident form in the limit $k\to 1$. In this limit the period of the non-commensurate structure tends to infinity ($L\to\infty$), and the solution (16) describes the propagation of a moving kink through another standing kink: the last term in (16) transforms into the well-known expression for a moving soliton

$$\delta u(x,t)=\pm 4\arctan\exp\left\{\left[x\pm\upsilon t-\gamma(x)\right]/\sqrt{1-\upsilon^2}\right\}, \qquad (17)$$

where $\gamma(x)$ is a localized function which describes the deformation of a kink during its propagation through the standing kink and depends on functions φ

and ρ. The polarity of the kink and the sign of its velocity depend on the sign of the parameter μ and the value of the parameter λ.

In conclusion, we studied analytically the nonlinear dynamics of the non-commensurate structure of a surface atomic layer with a spatially periodic ground state. The nonlinear excitations of the periodic soliton lattice (moving additional kinks) were investigated in the framework of the Frenkel-Kontorova model. We think that the nonlinear excitations of non-commensurate surface structure may be detected experimentally if the wave with frequency in the gap of the spectrum will be excited near the surface.

Acknowledgments

ASK thanks the INTAS for support under INTAS-99 Programme (Grant No.0167). IVG acknowledges the Consejo Superior de Investigaciones Científicas (CSIC) and ICMM, Madrid, Spain, for hospitality and support.

References

1. D.L. Mills and S.E. Trullinger, Phys. Rev. B **36**, 947 (1987).
2. O.A. Chubykalo, A.S. Kovalev, and O.V. Usatenko, Phys. Rev. B **47**, 3153 (1993).
3. A.B. Aceves and S. Wabnitz, Phys. Lett. A **141**, 37 (1989).
4. U. Harten, A.M. Lahee, J.P. Toennies, and Ch. Wöll, Phys. Rev. Lett. **54**, 2619 (1985).
5. M. Mansfield and R.J. Needs, J. Phys.: Condens. Matter **2**, 2361 (1990).
6. O.M. Braun and Yu. S. Kivshar, Phys. Rep. **306**, 1 (1998).
7. A.M. Kosevich and A.S. Kovalev, *Introduction to Nonlinear Physical Mechanics* (in Russian, Naukova Dumka, Kiev, 1989).
8. M.A. Sall', Teor. Mat. Fiz. **52**, 227 (1982).
9. A.M. Kossevich, *The Crystal Lattice: Phonons, Solitons, Dislocations* (Wiley-VCH, Berlin, 1999).

RAMAN SCATTERING AS AN EPIOPTIC PROBE FOR LOW DIMENSIONAL STRUCTURES

E. SPEISER[2], K. FLEISCHER[1], W. RICHTER[1,2]

[2] *INFM, Dipartimento di Fisica Università di Roma "Tor Vergata", Via della Ricerca Scientifica 1, I-00133 Roma, Italy*

[1] *Technische Universität Berlin, Institut für Festkörperphysik, Sekr. PN 6-1, Hardenbergstr. 36, D-10623 Berlin, Germany*

This paper reviews Raman scattering from the viewpoint of todays increased experimental possibilities which allow for more sophisticated experiments, specifically for applications in low dimensional solids. After a more general review of the basic aspects of Raman scattering we discuss several examples, dealing first of all with the main objective of inelastic light scattering the determination of vibrational frequencies. The Si(111):In-(4×1) surface, GaAsN layers within a GaAs matrix and Si nano rods serve as examples. The second extremely useful aspect of Raman scattering, is the resonance enhancement of the scattering cross section, if the photon energies match the energies of electronic states. This enhancement increases the sensitivity of the scattering process, an important fact especially for low dimensional systems where the total number of atoms in the scattering volume is much smaller than in the bulk. The second important aspect of resonant scattering is the information about the electronic states contained in the dependence of the scattering cross section on the incident photon energy.

1. Introduction

Since its discovery in 1928 [1, 2] Raman scattering has always strongly profited from developments in optical technology. The availability of gas lasers in the sixties simplified the measurements of opaque solids considerably and coined the term "laser Raman scattering". The development of optical multi-channel detectors in the last two decades (diode array, charged coupled devices – CCD), allowed for time resolved measurements and, especially the CCD, for increased sensitivity. Recently also solid state tunable lasers (Ti:Sapphire, laser diodes) have become available and replaced the problematic dye lasers as tunable light sources in resonant Raman scattering. In addition to these advances in light sources and detectors also new optical components (holographic optics) have become available which allow more sophisticated experiments and open up new fields for Raman scattering. Of special importance, are also improvements in the lateral spatial resolution of Raman scattering and should be emphasized here. By using microscopes or set-ups from scanning near field optical microscopy (SNOM) as input configuration to the monochromatic systems the spatial resolution is now well in the submicron range. Through these apparative developments sensitivity has been increased considerably and low dimensional structures, with many orders of magnitudes less atoms, have come now into the range of Raman scattering experiments.

Two-dimensional structures (surfaces, buried layers) of course have already been the object of Raman scattering for quite some time. More than twenty years ago electric field induced Raman scattering (EFIRS) from LO-phonons has been utilized to study the space charge layer at semiconductor surfaces [3–5]. Even more confined in the dimension normal to the surface microscopic surface phonons have been studied in the last decade. These studies have been also recently reviewed [6]. Detection of monolayers has been achieved [7–9]. Recently also Raman scattering measurements from quasi-1-dimensional (Q1D) structures (carbon nanotubes:[10, 11]; In chains on Si(111):[12]) as well as from 0-dimensional quantum dot structures (Q0D) [13] have been reported.

In this paper, after discussing the basics of Raman scattering we will concentrate on some recent low dimensional topics the authors have been involved with, but will point also out to other Raman work presently under discussion by giving appropriate references.

2. Fundamentals of Raman Spectroscopy

The basic features of Raman scattering from vibrational excitations are very similar and independent of the dimensionality of the system. This concerns the experimental techniques, the kind of selection rules, resonance phenomena and other properties. For this reason we will confine this section mainly to a short summary of these general properties as they have been discussed in several reviews dealing with bulk systems [14–16]. We will point out the modifications occurring in lower dimensional systems.

The main difference between Raman scattering from bulk systems and lower dimensional systems is that the latter have different electronic states and vibrational properties than the bulk. While these differences do not cause any principal limitations, the small number of atoms present in the lower dimensional systems turns out to be the main experimental obstacle. As a result the scattering intensities are low and the experiment needs to be carefully designed. Besides standard optical measures such as high aperture, optimized collecting optics for the scattered light between the sample preparation chamber and the monochromator, it turns out that the main advantage comes from the exploitation of cross sectional resonance enhancements (Resonance Raman Scattering). For this reason previous knowledge about the electronic band structure is extremely helpful and quite often the choice of photon energies is the decisive parameter for a successful experiment. Vice versa the experiment, which first of all is supposed to determine vibrational properties, allows also to gain information about the electronic states by using different laser lines for excitation.

2.1. *Energy and Wavevector Conservation*

Raman spectroscopy is understood as the spectral investigation of inelastically scattered light ($\hbar\omega = 1$ to $6\,\mathrm{eV}$) with energy transfers larger than approximately $1\,\mathrm{meV}$ ($8\,\mathrm{cm}^{-1}$). Grating monochromators (double or triple) with high resolution and, more importantly, with high contrast are usually employed for the spectral analysis. In the scattering process a certain amount of energy is gained or lost by an incident photon with energy $\hbar\omega_i$ (*incident*) in order to create or annihilate elementary excitations of the solid, usually phonons, resulting in a scattered photon of a different energy $\hbar\omega_s$ (*scattered*). The amount of energy transferred corresponds to the eigenenergy $\hbar\omega_j$ of the elementary excitation involved labeled by "j".

$$\hbar\omega_s = \hbar\omega_i \pm \hbar\omega_j \tag{1}$$

Here the "minus" sign stands for a phonon excitation (Stokes process) while the "plus" sign implies a phonon annihilation (anti-Stokes process).

The momentum $\vec{k}_j$ transferred to the vibrational excitation is related to the momentum of the incident $\vec{k}_i$ and scattered light $\vec{k}_s$ according to:

$$\hbar\vec{k}_s = \hbar\vec{k}_i \pm \hbar\vec{k}_j. \tag{2}$$

It is quite small in periodic structures when compared to the maximum possible quasi-momentum at the edge of the Brillouin Zone. Thus one usually assumes $\hbar\vec{k}_j = 0$. In low dimensional solids, when the periodicity is reduced in one or more directions, of course the wave vector becomes meaningless in those directions and one should discuss the phonons as confined modes.

2.2. *Scattering Intensity*

Raman scattering, involving two electromagnetic fields with different frequencies, may be discussed in terms of a higher order dielectric susceptibility [7, 15–17]. This has been termed transition susceptibility [18], a name which refers to the fact that, in addition to the interaction of the electromagnetic fields with the solid, the creation or annihilation of an elementary excitation is involved.

The standard linear dielectric susceptibility (tensor) $\tilde{\chi}(\omega, \vec{\mathrm{k}})$ describes the first order interaction (reflection, transmission, elastic scattering) of an electromagnetic wave $\vec{\mathrm{E}}(\omega, \vec{\mathrm{k}})$ with matter. Due to the small $\vec{\mathrm{k}}$-vector of visible light the explicit $\vec{\mathrm{k}}$-dependence can be neglected in most cases, and the induced polarization reads:

$$\vec{P}(\omega) = \epsilon_0 \tilde{\chi}(\omega) \vec{E}(\omega). \tag{3}$$

The generation of scattered light with frequency ω_s by incident light with frequency ω_i may be expressed using the transition dielectric susceptibility tensor $\tilde{\chi}(\omega_i, \omega_s)$ that connects the exciting electromagnetic field with frequency ω_i to the scattered field with frequency ω_s:

$$\vec{P}(\omega_s) = \epsilon_0 \tilde{\chi}(\omega_i, \omega_s) \vec{E}(\omega_i). \tag{4}$$

Here $\vec{P}(\omega_s)$ is the oscillating polarization which gives rise to the scattered light wave and $\vec{E}(\omega_i)$ is the oscillating electric field of the incident light wave.

The scattered intensity can finally be expressed as dipole radiation using this generalized dielectric susceptibility $\tilde{\chi}(\omega_i, \omega_s)$ which is also often called Raman tensor [7]:

$$I_s = I_i \frac{\omega_s^4 V}{(4\pi\epsilon\epsilon_0)^2 c_0^4} |\vec{e}_s \epsilon_0 \tilde{\chi}(\omega_i, \omega_s) \vec{e}_i|^2 \tag{5}$$

where I_i, I_s and $\vec{e}_i$, $\vec{e}_s$ denote the intensity and polarization, respectively of incident and scattered light.

The generalized susceptibility tensor can be related to the linear susceptibility in a quasistatic picture by a Taylor expansion in terms of the lattice deformation. Thereby, one assumes a modulation of the linear susceptibility proportional to generalized coordinates $\mathrm{Q_j}$ (normal coordinates), which correspond to the lattice deformation caused by the phonon excitation:

$$\begin{aligned}\chi_{\alpha,\beta}(\omega_i, \omega_s) = {} & \chi^o_{\alpha,\beta}(\omega_i) \\ & + \sum_j Q_j \cdot \left(\frac{\partial \chi_{\alpha,\beta}(\omega_i)}{\partial Q_j} \right) \\ & + \sum_{j,j'} Q_j Q_{j'} \cdot \frac{1}{2} \left(\frac{\partial^2 \chi_{\alpha,\beta}(\omega_i)}{\partial Q_j \, \partial Q_{j'}} \right) \\ & + \ldots\end{aligned} \tag{6}$$

In a microscopic quantum mechanical approach the light scattering process may be described using time-dependent perturbation theory [19]. The dominant term amounts to [20]:

$$\chi_{\alpha,\beta}(\omega_i,\omega_s) = \frac{e^2}{m_0^2 \cdot \omega_s^2 \cdot V} \sum_{e,e'} \frac{\langle 0|p_\alpha|e'\rangle\langle e'|H_{E-L}|e\rangle\langle e|p_\beta|0\rangle}{(E_{e'} - \hbar\omega_s)(E_e - \hbar\omega_i)} \tag{7}$$

where m_0 is the electron mass, V the scattering volume, p_α, p_β the Cartesian components of the momentum operators, E_e, $E_{e'}$ the energies of the excited electron-hole pair states and H_{E-L} the electron-phonon interaction Hamiltonian.

Equ. (7) includes the transition from the ground state $|0\rangle$ to an excited electronic state $|e\rangle$ (photon absorption), scattering of the generated electron-hole pair into another state $|e'\rangle$ via electron-lattice interaction, and finally the transition back to the electronic ground state $|0\rangle$ under photon emission.

If only two bands (states) are involved in the scattering process, equ. (7) can be rewritten also as a frequency derivative of the electric susceptibility in the form [16]:

$$\chi_{\alpha,\beta}(\omega_i,\omega_s) = D_j \cdot \frac{d\chi_{\alpha,\beta}(\omega)}{d(\hbar\omega)} \tag{8}$$

where D_j is the corresponding electron-phonon interaction matrix element for the electronic state and the vibrational excitation under question. The derivation assumes that this matrix element is constant for all electronic transitions involved. Only under such an assumption it can be taken in front of the summation of equ. (7) and the remaining sum then just represents the derivative of the susceptibility with respect to energy. For that reason Raman scattering can be viewed also as a kind of modulation spectroscopy [20].

The relationships (7) and (8) can be utilized to calculate the Raman scattering cross section. The transition susceptibility $\tilde{\chi}(\omega_i,\omega_s)$ is obtained by the derivative of the linear susceptibility $\tilde{\chi}(\omega)$ with respect to energy (equ. 8) or, more generally, with respect to the lattice deformation (equ. 6). The Raman cross section can consequently be obtained via band structure calculations, by taking the difference of the dielectric susceptibilities for the equilibrium lattice and that obtained after a shift of atomic positions according to the phonon normal coordinates [15, 16, 21]. The energy derivative is also quite useful in situations where experimental data for the susceptibility are available. This is quite often the case since dielectric functions can be determined by ellipsometric measurements.

The electronic properties that may be influenced by a phonon, are the eigenenergies and eigenfunctions of the electronic states. If the mechanical lattice deformation caused by the phonon is the only microscopic origin of the modulation of electronic properties, the interaction mechanism is called deformation potential scattering. This has been intensively discussed for resonant Raman scattering with photon energies at different energy gaps in III-V semiconductors [16].

An additional interaction mechanism may be involved for the longitudinal vibration modes of IR-active bulk optical phonons, i.e. phonons for which the symmetry of their eigenvectors allows a dipole moment per unit cell [18, 22]. The longitudinal component of such phonons generates, for small wavevectors, a macroscopic, long range electric field in the three dimensional case. Besides acting as an additional restoring force and thus leading to an increased frequency of the longitudinal optical phonon with respect to the transverse partner ($\omega_{LO} > \omega_{TO}$), this field may also give rise to an additional light scattering mechanism since it may interact directly with the electrons. Especially when the electronic excitations involve excitonic contributions the electronic wavefunctions are strongly modified by the electric field. This interaction mechanism is called *Fröhlich interaction* [23]. While being the dominant mechanism for bulk phonon scattering at many gaps of III–V- and II–VI-semiconductors semiconductors with strong excitonic correlation between electron and hole, in lower than three dimensions the corresponding macroscopic electric field will go to zero in the small wavevector limit [24]. Thus the Fröhlich interaction should be of little or no importance in low dimensions. Indeed up to now experiments have not indicated such a scattering contribution, which in the bulk case manifests itself by specific selection rules and strong resonances with states involving excitonic contributions.

Apart from the ω_s^4-dependence corresponding to the dipole radiation (equ. (5)), the Raman scattering cross section will show a pronounced dependence on the energy of the exciting photons. Maxima in the Raman cross section will occur for photon energies matching critical points of the electronic band structure (equ. (4), (6), (7)). This condition, called *Resonant Raman Scattering*, has been intensively exploited for bulk phonon scattering [15, 16] and plays a similar role for low dimensional systems (Ref.:this work, [7]).

The enhancement of the Raman cross section under resonant conditions is of crucial importance for Raman experiments in general but even more for low dimensional systems with their reduced total number of atoms. Besides this enhancement of scattering intensity, even more important, may be the determination of the spectral dependence of the Raman cross section. This will additionally allow to analyze selectively the electronic properties of the system where the scattered light is generated.

Experimentally this can be accomplished by varying the exciting photon energy (different laser lines or tunable laser), monitoring quantitatively the Raman scattering intensity and determining the scattering efficiency derived from the differential cross section, which is proportional to the transition susceptibility (Raman tensor) and which has the dimension of an inverse length [15]:

$$S_{\alpha,\beta} = \frac{dI_{\alpha}^{s}}{LI_{\beta}^{i}d\Omega} = \frac{|\tilde{\chi}_{\alpha,\beta}(\omega_i,\omega_s)|^2\omega_s^4 V}{c_0^4} \tag{9}$$

where L is the length of the total scattering volume in the direction of the propagation of the incident light, Ω is the solid angle under which the scattered light is collected, and α, β denote the polarization of incident and scattered light. In bulk phonon scattering L will be often given by the penetration depth of the light. In case the sample is transparent L corresponds to the total thickness and similarly in low dimensional system L will be mostly fixed by the small volume of the structure. In the case of surfaces L is given by the thickness of the "surface", e.g. the localization depth of the surface phonon which amounts to a few monolayers in general. It should be noted that such quantitative measurement with different laser lines require some experimental efforts since not only the physical parameter of the structure studied but the experimental equipment as well is usually strongly dependent on frequency [15, 16].

In the susceptibility approach the Raman resonance is reflected in an enhancement of the partial derivatives of the linear susceptibility (equ. (5),(6)) at critical points of the electronic band structure. The validity of the susceptibility model can thus be experimentally verified by comparing the experimental result to the derivative of the susceptibility. This in fact has been done for several bulk semiconductors such as GaAs [25], InP [26] and Si [27]. The susceptibility approach should be also applicable to low dimensional systems. Indications for the validity in quasi-2-D systems have been presented in [6].

Another important feature of Raman scattering is the existence of selection rules. They describe the dependence of the scattering intensity on the polarizations of incident and scattered electric fields. They are derived by group theoretical analysis of the Raman tensor $\tilde{\chi}$ under the symmetry operations of the point group describing the system. Such an analysis has been performed for all irreducible representations of the 32 crystallographic point groups. The non-zero components of these tensors can be found, for example, in [15, 28]. For lower dimensional systems Raman tensors have not been generally listed, to our knowledge. However, since the symmetry groups for lower-dimensional systems form a subset of the corresponding three-dimensional ones, it is easy in specific cases to refer to the appropriate three-dimensional Raman tensors.

Since the lower-dimensional groups contain less symmetry operations than the three-dimensional groups it follows that the symmetry groups, and correspondingly the number of irreducible representations, are smaller. Therefore, in lower dimensions Raman selections rules for vibrations pose fewer restrictions and usually the vibrational modes will be Raman active.

Experimentally, certain tensor components are selected by choosing the polarization directions $\vec{e}_i$ and $\vec{e}_s$ of incident and scattered light along appropriate crystal symmetry directions. The experimental configuration is given in the so-called Porto-notation as $\vec{k_i}(\vec{e_i}; \vec{e_s})\vec{k_s}$ [15, 16].

The main topics to be discussed in the following sections are the determination of vibrational frequencies. This is especially interesting in dimensionally reduced systems and should be studied as a function of system size. The line width of vibrational excitations is another interesting feature in low dimensions. It should give information on the coupling to the surrounding system (damping). Such measurements however have yet to be exploited. The scattering intensity is the other most important source of information which can be obtained from Raman scattering. Section 4 will deal this topic.

3. Determination of Frequencies

The determination of phonon frequencies is the most obvious and straightforward result from Raman scattering experiments. The comparison of measured values and those calculated from lattice dynamical models gives insight into vibrational properties and bonding structure of the system under investigation. For bulk materials such comparisons, also performed under the influence of external parameters like temperature or stress, have been performed intensively in the last decades. In comparison to lattice vibrational information from neutron or electron scattering the light scattering results have a much higher accuracy and frequency resolution and thus are very helpful in complementing the information on the lattice dynamics.

3.1. *Surface Phonons*

Surfaces (interfaces) support new phonon modes with frequencies different from the bulk phonons. They are confined to the surface (interface) e.g. their amplitudes decay exponentially into the bulk [29, 30] and only wave vectors parallel to the surface are allowed. They are termed macroscopic if their penetration into the bulk is of the order of their in-plane wavelength (Rayleigh-, Fuchs-Kliewer-modes) or microscopic in case they are confined to the surface within a few interatomic distances [31], [30].

The classical analytical surface tools for surface excitations are HAS (helium atom scattering) [32] and HREELS (high resolution electron energy loss spectroscopy) [33] because of their small penetration (a few atomic monolayers) into the solid. In comparison with these two methods Raman scattering has the disadvantage of being less surface sensitive by penetration depth (100 monolayers as compared to 2 - 3 in HAS, HREELS) but has the advantage of a considerably higher spectral resolution in a wider energy range. With Raman spectroscopy surface phonon modes were found on clean InP(110) surfaces [34] and for adsorbate terminated surfaces such as InP(110):Sb [35], InAs(110):Sb [36] or Si(111):H [37]. Due to the large penetration depth of the light the spectra not only show the surface phonon modes (confined to a few monolayers), but also the bulk phonon modes, usually also with a much stronger intensity than the surface modes. Thus it is of advantage to study materials with a larger gap between acoustical and optical phonons. This reduces the possible interaction between surface and bulk phonons in the surface region but reduces also the spectral overlap in the scattering spectrum. Since in general one can also exploit resonance enhancement of the surface phonons with surface electronic states the intensity ratio of surface to bulk phonon scattering intensities is usually much higher than expected from the penetration depth argument alone.

Surface modes are identified either from the fact that they are occurring just with the adsorbate on the surface or, in the case of clean surfaces, because they disappear when an adsorbate (e.g. oxygen) destroys the reconstructional order of the surface.

3.2. *Si(111):In-(4×1) surface phonons*

The Si(111):In-(4×1) surface has gained considerably interest in recent years because 1D rows of In atoms are formed on the surface and at low temperatures ($<$ 120 K) the (4×1) surface undergoes a phase transition to a (4×2)/(8×2) reconstruction which is still under study [38–40]. The Si(111):In-(4×1) surface is obtained by depositing approximately 1 mono-

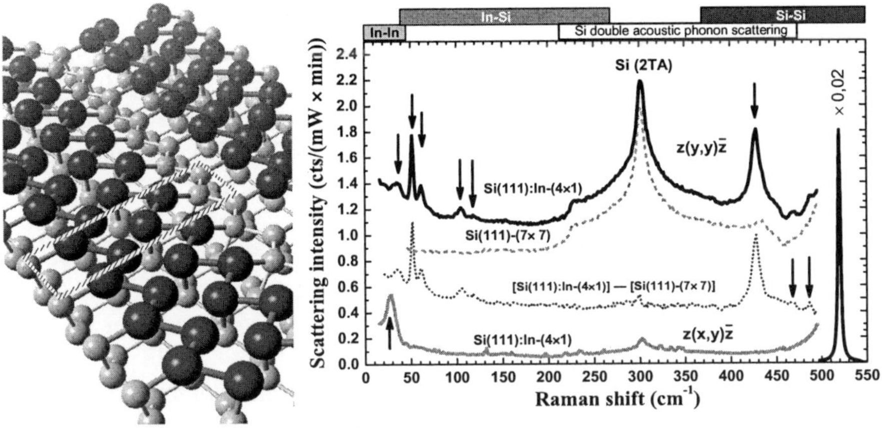

(a) (b)

Figure 1. Surface structure and Raman spectra of the Si(111):In-(4×1) surface. (a) Reconstruction model of the Si(111):In-(4×1) surface according to [41, 42]. The dark and grey circles represent In and Si atoms, respectively. The (111) direction is perpendicular to the plane shown, Indium chains are oriented along the $(\bar{1}\bar{1}0)$ direction. The unit cell is indicated by the dashed line. (b) Raman spectra of the Si(111):In-(4×1) surface measured with the 647 nm Krypton Laser in two different scattering geometries. The full black curve shows measurements with incoming and measured polarization parallel to the indium chains, while in the case of the full grey curve the incoming polarization was normal to the chains. The dashed line corresponds to the spectrum of the (In free) Si(111)-(7×7) surface, while the dotted curve shows the differences between spectra of Si(111):In-(4×1) and Si(111)-(7×7) surfaces. All surface vibrational modes of Si(111):In-(4×1) are marked with arrows.

layer of In on a Si(111) surface. In the by now most accepted structural model of this surface the In forms double zig-zag chains along the [110] direction on the silicon surface (Fig. 1a)[41, 42]. This surface structure could be possibly also treated as an assembly of quasi 1-d structures on Si(111).

Until recently not much was known with respect to the vibrational modes of these surfaces. In the calculations the number of modes depends of course on how many atoms are taken within the surface Brillouin zone e.g. how many bulk layers are considered. The first ab-initio calculations considered the 6 surface atoms (4 In + 2 Si) plus 8 Si atoms from the layer below thus altogether 14 atoms. Group theory with 14 atoms per unit cell predicts 42 (=14 ×3) modes of which 28 transform with A' and 14 with A'' symmetry [43]. Until recently only HREELS data were available which showed just one surface phonon mode at 484 cm^{-1}[44]. This high frequency mode, which occurs close to the TO frequency of silicon, probably involves mainly Si atomic displacements. There should be however, additional modes involving more dominantly In-displacements, with much lower frequencies, since the atomic mass ratio $m_{In}/m_{Si} = 4.08$ is quite large. Indeed the Raman scattering spectra shown in Fig. 1b for two scattering configurations display a number of additional peaks. All modes marked with arrows are surface vibrational modes. This is verified by comparison with spectra from the clean Si(111)-(7×7) surface before indium deposition, which exhibits only structures originating from Si phonon modes. By subtracting the spectra of the clean surface from the one of the Si(111):In-(4×1) (as shown in Fig.1b) only the surface vibrations of the latter remain.

Stimulated by these Raman measurements calculations of the vibrational properties of the Si(111)-In:(4×1) surface were recently performed by an ab-initio calculation (DFT-GGA) [43]. Vibrational force constants were derived by displacing each of the 14 topmost atoms of the slab (3 layers) by 0.01Å along each coordinates. The eigenvalues of the resulting 42×42 dynamical matrix can then be calculated leading to 42 vibrational

symmetry	measured (cm^{-1})	calc. (cm^{-1}) [43]
A''	28 ± 1	**15,21**
	–	51
	–	79
	–	81, 109
	–	421
A'	31 ± 1	**31**
	36 ± 2	**42**
	51 ± 0.6	48, **54**
	62 ± 1.3	68
	72 ± 3	78
	105†	93, 104
	116†	109
	(142)	129
	(150)	140
	–	201, 211, 245
	–	254, 274, 293
	–	303, 314
	–	384, 394, 401
	428 ± 1	**411**, 416, 430
	467 ± 1.5	444, 458
	487* ± 3	469

Table 1. Comparison of measured Si(111)-In:(4×1) phonon-mode frequencies with calculations. The phonon frequencies of all measured modes in $A'(z(y,y)\bar{z})$ and A'' $(z(x,y)\bar{z})$ are given. Following [43] the calculated modes set in bold denote modes with considerable contributions of the first layers. The assignment to measured modes is based only on the calculated energy.

* – also seen in HREELS [44].

† – two phonon process

modes – 28 in A' and 14 in A'' symmetry (point group C_S). The latter ones can be seen in Raman spectra with crossed incoming and scattered polarization $z(x,y)\bar{z}$, while the A' modes can be seen for parallel polarization $z(y,y)\bar{z}$. The experimental and calculated frequency values are compared in Table 1. Still an assignment of the 42 calculated modes to the 11 measured ones is not straight forward. One likely reason is that a number of modes probably cannot experimentally be observed because of the strong Si second order scattering in the 200 - 400 cm^{-1} region. Secondly, for a number of measured peaks there is the possibility that more than one mode contributes since the energetical differences of some calculated modes are small. Even other peaks such as the ones measured at 105 and 116 cm^{-1} originate from two phonon scattering processes and therefore cannot directly compared to a calculated frequency. From the calculational point of view there may be also possible deviation between calculated energies and measured ones due to the small slab size in the calculations. In summary the large number of calculated modes (due to the large basis) makes the assignment difficult. If one limits the discussion to calculated modes with displacement patterns limited to movements of the top layer atoms (In-In or Si-Si vibrations within the chain structure) as set in bold in Tab. 1, the agreement is very good.

3.3. *Vibrations in buried layers*

In the foregoing the large penetration depth of the light was discussed under the view point of surface sensitivity and seen as an disadvantage in comparison to HREELS and HAS. Because of the larger scattering volume extending 50 up to 100 ML the bulk phonons will dominate the Raman spectra. However, the larger penetration depth on the other hand allows also to reach regions below the surface which are not accessible by the other two techniques. Thus Raman scattering can give information on buried layers or interfaces deeply below the surface. The following example of GaAsN layers within a GaAs matrix demonstrates the detection of such vibrations.

GaAsN (as well as GaInAsN) with a low nitrogen content is a technologically very interesting ternary material since the addition of N into GaAs reduces the band gap and makes it a potential candidate for IR-optoelectronic and solar cell applications [45, 46]. It has, moreover, the advantage that one can use the established GaAs technology. The main experimental problem is to introduce sufficient quantities of nitrogen into GaAs and also to determine the amount and distribution incorporated.

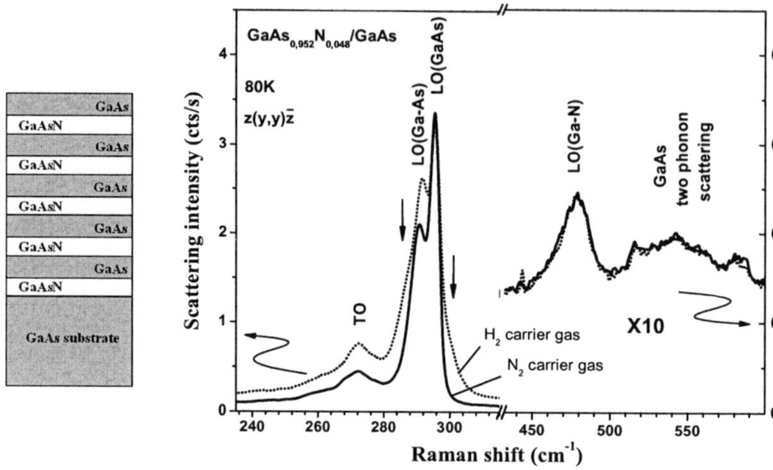

(a) (b)

Figure 2. Determination of vibrational properties in "buried" structures. (a) Schematic cross section of MOVPE grown GaAsN/GaAs layer structure with ca. 5.5 nm GaAsN and ca. 9 nm GaAs layer thicknesses. (b) Raman spectra of two similar GaAsN/GaAs layer structures grown with N_2 (black curve) and H_2 (grey curve) as carrier gas. Additional phonon lines appear at the arrow marked positions in the spectra of the sample grown in H_2 carrier gas (see Fig. 3 for details).

Epitaxial growth of GaAsN has been done by either MBE [47] or MOCVD [48]. In Fig. 2 we show Raman spectra of two different GaAsN/GaAs layer structures grown by MOVPE with two different carrier gases (N_2 and H_2) but with nearly identical nitrogen content (4.8 % as determined by X-ray diffraction). Besides the LO(GaAs), new modes labeled LO(Ga-N) and LO(Ga-As), appear in the Raman spectra. While the LO(GaAs) is generated within the GaAs spacing layers or/and the substrate, the LO(Ga-N) has been shown by ion implantation of nitrogen isotopes and IR absorption measurements as local vibrations of Ga-N bonds [49]. In the presence of such Ga-N bonds, which are shorter than the Ga-As bonds, also the lower frequency LO(Ga-As) can be explained through dilatation of the surrounding GaAs leading to a lower frequency then the LO(GaAs) [50, 51]. The occurrence of the two modes thus shows that the nitrogen is not homogeneously distributed in the GaAsN layers, and formation of Ga-N clusters occurs.

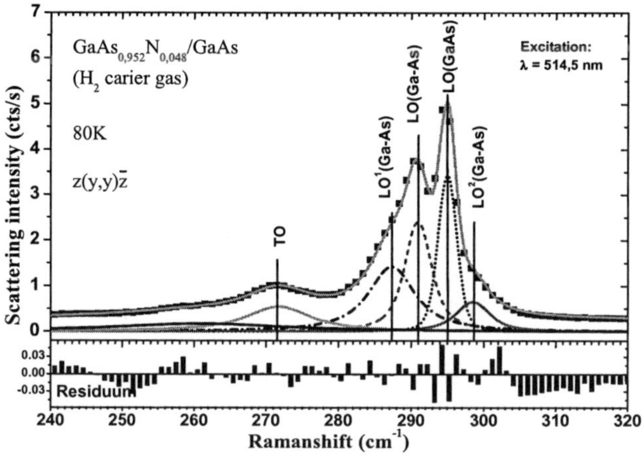

Figure 3. Line analysis of a Raman spectra from MOVPE grown GaAsN/GaAs with H_2 carrier gas (already shown in Fig. 2). Compared to another spectra additional lines appear between TO and LO(Ga-As) and above LO(GaAs). The residuum for the shown fit (difference between measured data and fit function) is placed in the lower part of the graph.

The appearance of the TO phonon which, in the scattering configuration shown, is forbidden by zincblende symmetry selection rules proves similar, that the inhomogeneous inclusion of nitrogen disturbs additionally the zincblende symmetry and leads to a disorder in the sample.

In spite of very similar structure and the nearly identical stoichiometry of the two samples the Raman spectra in Fig. 2 show a clear difference at the positions marked by the vertical arrows. We therefore undertook a line fit analysis of the H_2-carrier-gas-sample Raman spectrum (upper spectrum, dotted curve) with a Voigt line shape. The mathematical procedure results from the convolution of a Lorentzian line shape (phonon) with the instrumental Gaussian broadening (monochromator) and is given by [52]:

$$V(\omega) = A_0 + A^2 \frac{2\ln 2}{\pi^{\frac{3}{2}}} \frac{\sigma_L}{\sigma_G^2} \int_{-\infty}^{\infty} \frac{\exp -\omega'^2}{\left(\sqrt{\ln 2}\frac{\sigma_L}{\sigma_G}\right)^2 + \left(\sqrt{4\ln 2}\frac{\omega-\omega_0}{\sigma_G} - \omega'\right)^2} d\omega'. \quad (10)$$

with σ_G, σ_L being the FWHM of the Gaussian and Lorentzian lines and ω_0 the peak position. This procedure is well known also in many other spectral analysis techniques. The detailed result of the fit is shown in Fig. 3. The accuracy of the fit can be judged from the residuum (shown in the bottom of Fig. 3). Since this has no peak structure and only very small values (below

1 %) the two additional phonon lines LO^1(Ga-As), LO^2(Ga-As) appearing at the positions previously marked by arrows in Fig. 2 are significant.

The appearance of additional lines in this sample with the same stoichiometry might look surprising at the first glance. However, as discussed before in connection with LO(Ga-As) and LO(Ga-N), this probably originates from different distributions of Ga-N clusters. Thus, even the average macroscopic concentration is the same, the inhomogeneities are different and create on a mesoscopic scale additional LO(Ga-As) peaks from differently dilated GaAs regions (left side of LO(GaAs)) or compressed GaAs regions very near the Ga-N clusters (right side of LO(GaAs)). Moreover, the higher intensity of the symmetry forbidden TO in the H_2-grown-sample confirms the larger disorder. In conclusion for the two MOVPE samples grown under different carrier gases it can be stated that the H_2-carrier gas sample contains more inhomogeneities than the N_2-carrier gas sample. A possible explanation for this are different growth temperatures at the surface caused by the different thermal conductivities of the two carrier gases [53].

3.4. *Nano-rods*

The previous examples discussed, demonstrate the ability of Raman spectroscopy to determine the vibrational properties of low dimensional structures. These structures however were supported by the bulk material (thin layers, wires on a surface) and thus interactions with the bulk vibrational modes have to be considered. Nowadays also free standing low dimensional structures like nanorods or nanotubes can be produced and offer perhaps cleaner cases for studying confinement in low dimensional structures. The goal is consequently to study the frequency and linewidth of the phonons as a function of the size and possibly also the shape of the nano particles and comparison to the bulk values. Simple semi-empirical models, describing the confinement through a Gaussian wave vector distribution, have been already developed more than 20 years ago for the description of Raman lines of nano crystalline Si [54] and is widely used in order to describe nano particles Raman line shape [55–58]. While the theoretical approach to Raman frequencies from the bond polarizability model point of view was discussed for some nano structures [59–62], unfortunately, to our knowledge no general ab-initio treatment is available yet.

One of the experimental problems in Raman scattering from free standing nano-structures is the thermal management of the incident laser power within the individual nanostructure since the thermal conductance via the

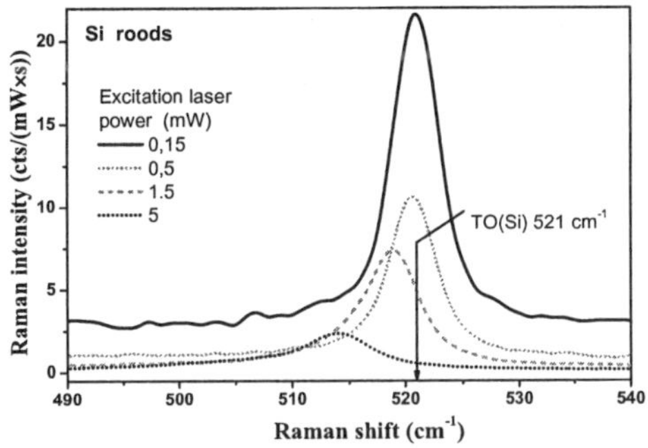

(a) (b)

Figure 4. Laser power dependent Raman spectra of Si wires. (a) SEM picture of Si wires prepared by CVD on Si (111) substrate. The average diameter of the wires is in the range of some hundred nanometers. (b) Micro Raman spectra of sample shown in (a) at different excitation laser power in the range from 5 to 0.15 mW. Notice that the excitation power independent intensity cts/(mW× s) is plotted. The Raman mode shifts with decreasing power to higher wavenumbers until it remains at the position of crystalline Si (arrow marked position at 521 cm^{-1}). The line asymmetry observed for higher excitation power disappears at lower power.

surrounding material is largely reduced. Thus the exciting laser power easily may create non-equilibrium conditions. This situation occurs especially in absorbing materials like Si-nanowires. Such conditions have been already discussed for Raman scattering in Si bulk under the aspect of laser crystallization [63]. The temperature increase induced by the laser depends on many parameters (laser power, thermal conductivity, heat capacity of the sample, thermal radiation) and is difficult to control. Moreover, even with a micro Raman setup the focus ($\o \sim 1\mu m$) on a sample as shown in Fig. 4a will cover an area containing several Si-nanowires which possibly might have also different geometries.

The experimental solution to this problem is to measure spectra with different laser powers as shown in Fig. 4b where the scattering intensity has been normalized with the incident laser power. The extrapolation to zero power gives the geometrical confinement effect. The Si-nanowire phonon show a strong shift to lower wavenumbers and an asymmetric broadening with increasing laser power similar as those reported in Ref. [64]. This asymmetry is usually discussed in terms of the geometrical boundaries in the low dimensional particles which as a consequence lead to an extended range of k vectors in reciprocal space (relaxation of k conservation). The

Raman line shape is then obtained by the weighted integration across the phonon dispersion curves [54]. However, the line asymmetry in Fig 4(b) disappears for low excitation power. This shows that the asymmetry of nano particles Raman lines can not always be described by simple relaxation of k vector conservation. The strong asymmetric broadening was also observed for Si bulk at high temperature [65]. Thus for nano particles it could possibly result from the anharmonicity of the interatomic potential. The temperature increase estimated from the frequency shift of bulk Si phonon data at high temperature [66] amounts to ca. 500 K for excitation laser power of 5 mW in Fig. 4b, but also higher temperatures up to ca. 1400 K where observed at higher excitation power. Thus these wires offer the interesting possibility to study anharmonic effects. The extrapolated "zero" power spectrum on the other hand should give the confinement effects. However in this example where the nanowires have thicknesses of 100 nm or more such confinement effects are probably not very significant. Nanowires with smaller thickness (<10 nm) should be used to study confinement effects in quasi-1D structures. They offer, moreover, the excellent opportunity to study anharmonic effects in nano-structures.

4. Resonance of the Raman cross section

As discussed already in Sect. (2) and can be seen in equ. (7) the Raman cross section will show strong enhancements whenever the photon energies match the energies of electronic transitions. Close to resonance ($E_{e'} - E_e < \hbar\omega_j$) this can lead to "in" and "out" resonances with the incident or the scattered radiation or in special cases even to double resonance ([67–69]). Such phenomena are observed when either the incident photons or the scattered photons or both are in resonance e.g. the corresponding energy denominators in equ. (7) become small. Here we will not deal with such situation of extreme resonance but more with situations where the photon energies are near but still sufficiently more than the vibrational energy away from the relevant electronic states.

This near-resonant situation has several advantages. First of all experimentally it is much easier to perform, since it is difficult to match from the exact resonance condition for a given material system with the still limited choice of available laser lines. Nevertheless, one usually can still take advantage of an considerable enhancement of the scattering intensity. This allows quite often to study processes which, in non-resonant conditions, would be to weak to be detected or to be studied quantitatively. Moreover, the dependence of the cross section over a larger range of photon energies can be exploited to find out which energy states are involved in the resonance. A priori this is in general not known in nano-structured solids. We will give two examples in the next subsections.

4.1. *Analysis of ZnO based nanostructures*

ZnO surfaces are an important low dimensional structure already since decades. Their conductivity is known to be extremely sensitive to gas adsorption [70]. Lately it has become even more important because of its large bandgap (3.3 eV) as a material for "blue" optoelectronic applications. Moreover, it is possible to grow epitaxial nanorods which can act either as single light guides or in a ordered fashion as photonic crystals [71]. Combined with the sensitivity to gas adsorption this could make enhanced sensor applications possible.

One necessary task in order to proceed with these applications is the analysis of the nanorods with respect to their reduced dimensions. Phonons detected by Raman scattering are one of the possible analytical tools in order to test the quasi 1-dimensional system. Thereby conventional Raman scattering techniques provide an average information from many nano structures within the excitation area. Excitation areas for common RS setups vary from several square micrometers for microscope supported RS to some square millimeters for simple focusing of the laser beam (macro-RS). High lateral resolution RS techniques with excitation areas on nanometer scale, presently under development, open the possibility to investigate vibrational properties of single nano structures [72]. However, to cope with the reduced amount of ZnO material, provided by a nanorod, resonance conditions are necessary.

ZnO shows a strong resonance behavior as can be seen in Fig. 5a with the example of two Raman spectra of bulk ZnO taken with different laser lines. They are in general agreement with the data which have been published in the past [73–76]. The spectrum with the 647.1 nm (= 1.95 eV) line of a Krypton laser is non-resonant with respect to the band gap of ZnO (3.3 eV) while the 406.7 nm (= 3.05 eV) line is near resonant. Clearly the enhanced scattering intensities with the latter excitation can be seen in Fig. 5. Under such resonant conditions it should be possible to observe phonon spectra from single, isolated nanorods.

By using near resonant excitation (406.7 nm) Raman measurements of ZnO nanorods grown by CVD on Si(100) substrate were possible (Fig. 5b). In spite of the low ZnO nanorod density on the substrate surface (ca. 5 nanorods per μm^2, each one approximately 40 nm thick and 300 nm long) and thereby small scattering volume, identification of ZnO vibrational modes (E_2, 2LO) is possible. No size related frequency shifts, however,

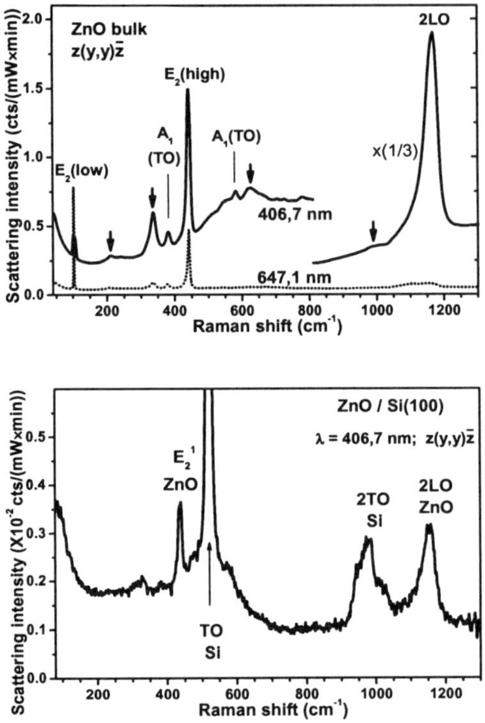

(a)

(b)

Figure 5. Raman spectra of ZnO: (a) bulk ZnO, for two different laser lines. With 647.1nm excitation a non-resonant Raman spectrum is obtained. The 406.7 nm spectrum is **near resonant** showing a larger scattering intensity and a number of second order phonon peaks, 2LO and indicated by arrows. The indexed capital letters refer to the irreducible representations of the point group C_{6v} of wurtzite ZnO.; (b) ZnO nano rods on a Si(100) substrate, luminescence background has been subtracted. Due to near resonance conditions enhanced E_2^1 and 2LO modes are observed.

were detected within the accuracy of this measurement ($\approx$ 1 cm^{-1}). The size of the the ZnO nano rods is probably still not small enough to generate observable phonon shifts as reported in the literature [77].

4.2. *In/Si(111)*

Raman spectra of In-wires on Si(111) were already shown and discussed in Sect. 3.2, Fig. 1. Here we discuss such Raman spectra taken with different laser lines (Fig. 6). Clearly, one recognizes the different scattering intensities

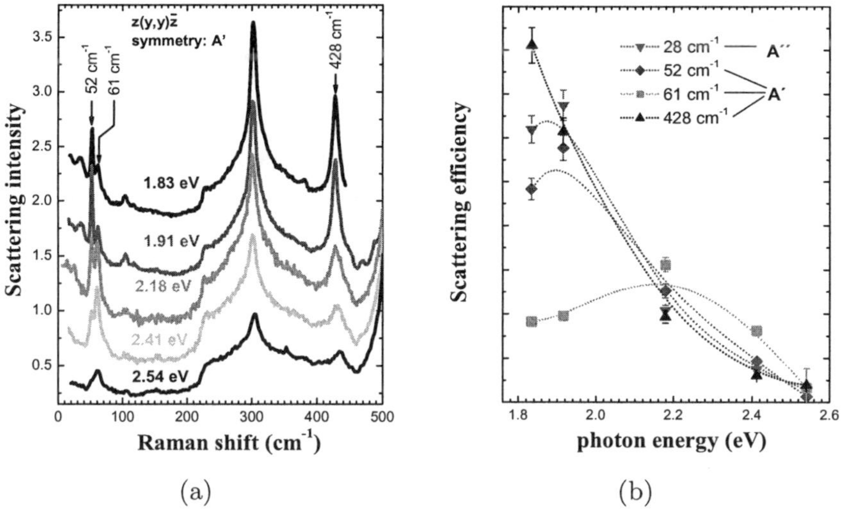

Figure 6. Resonance Raman scattering from Si(111)-In:(4×1). (a) Raman spectra normalized with the TO phonon intensity for various incident photon energies. The polarization vector y is [110] e.g. perpendicular to direction of the Indium chains. The spectra are displaced vertically for better visibility. (b) relative scattering efficiencies plotted versus incident photon energy for different phonons. The efficiencies of A'' mode at $28\,\mathrm{cm}^{-1}$ were obtained from a different set of Raman spectra taken in z(xy)-z polarization.

when the excitation photon energy is changed. For a quantitative discussion the scattering intensities should be evaluated in terms of the scattering efficiencies, equ. (9) as a function of laser photon energy in order to eliminate all experimental parameters depending on photon energy.

In principle absolute values of the scattering efficiency can be obtained in that way. However, the whole procedure is very cumbersome and highly inaccurate, because of the different photometric measurements involved. This concerns first of all the incident power and the scattered power which often differ by up to 10 orders of magnitude and have to be measured with different detectors. More convenient is therefore if one can normalize in the same experiment with the scattering of a phonon whose scattering cross section is known (eg. substrate phonon) and eliminate in such a way the laser power dependence. In the case of the In/Si(111) surface the TO phonon of bulk silicon can be taken for that purpose

$$\frac{I_{\text{surf}}}{I_{Si,TO}} \cdot S_{Si,TO} \sim S_{\text{surf}} \tag{11}$$

where S_{surf} then gives a quantity proportional to the desired Raman scattering efficiency of the surface modes.

Fig. 6 compares spectra with normalized intensities $I/I_{Si,TO}$ and amplitudes of some modes corrected for penetration depth and the resonance of the silicon bulk mode.

By analyzing the corrected spectra, the Raman susceptibilities for all modes can be plotted versus the incident photon energy (Fig. 6b) for the various modes. It can be easily seen that most surface phonon modes are resonant around 2 eV – the same energy where also the strongest anisotropy in the reflectance of the surface occurs [78, 79]. Only the mode at 62 cm^{-1} shows a different resonance at 2.2 eV. This different behavior must be related to this particular A' -phonon which possibly couples to other electronic states than those modes resonantly enhanced in the red spectral region. To understand this in detail more theoretical information on the electronic states, particularly their localization on certain atoms and a comparison to the individual displacement patterns of the phonon modes must be available.

5. Conclusions

We have discussed several examples of low dimensional structures where phonon parameters could be extracted by Raman scattering. Thus from the experimental side one can be quite optimistic about the role of Raman scattering for low-dimensional physics. What needs to be technologically improved are the high spatial resolution techniques. Theoretically the situation is also rather unsatisfying at the moment. Calculations for interpretation of the observed (or not observed) frequency shifts with respect to materials and geometry are urgently needed.

Acknowlegments

The authors are grateful to Maurizio De Crescenzi for many discussions and the Si-nanorods sample. We thank Florian Poser for the MOVPE growth of the multi layer GaAsN/GaAs sample. Financial assistance through COFIN 2004 (protocol Nr. 2004028932_003) is acknowledged.

References

1. C. V. Raman, *Ind. J. Phys.* **2**, 387 (1928).
2. C. V. Raman, *Nature* **121**, 619 (1928).
3. A. Pinczuk and E. Burnstein, *Phys. Rev. Lett.* **21**, 1073 (1968).
4. N. Esser, M. Reckzügel, R. Srama, U. Resch, D. R. T. Zahn, W. Richter, C. Stephens, and M. Hünermann, *J. Vac. Sci. Technol. B* **8**, 680 (1990).

5. N. Esser, W. Richter, U. Resch-Esser, P. Chiaradia, C. Goletti, and L. Moretti, *Phil. Mag.* **B 70**, 507 (1994).
6. N. Esser and W. Richter, *Raman Scattering from surface Phonons in Topics in Applied Physics **Vol. 76**, Light Scattering in Solids VIII, ed. by M. Cardona and G. Güntherodt*, Springer, Berlin, Heidelberg, New York, 2000.
7. N. Esser and J. Geurts, *Optical Characterization of Epitaxial Semiconductor Layers, Ed.: G. Bauer, W. Richter*, Springer, Berlin, Heidelberg, New York, 1996.
8. N. Esser, K. Hinrichs, J. R. Power, and W. Richter, *Surf. Sci.* **427-428**, 44 (1999).
9. J. G. M. Hünermann and W. Richter, *Phys. Rev. Lett.* **66**, 640 (1991).
10. M. Milnera, J. Krti, M. Hulman, and H. Kuzmany, *Phys. Rev. Lett.* **84**, 1324 (2000).
11. J. Maultzsch, S. Reich, and C. Thomsen, *Phys. Rev.* **65**, 233402 (2002).
12. K. Fleischer, S. Chandola, N. Esser, W. Richter, and J. F. McGilp, *Phys. Rev.* **67**, 235318 (2003).
13. B. Bennett, B. V. Shanabrook, and R. Magno, *Appl. Phys. Lett.* **68**, 958 (1996).
14. W. Hayes and R. Loudon, *Light Scattering in Solids*, J. Wiley and Sons, New York, 1978.
15. M. Cardona, *Topics in Applied Physics **Vol. 50**, Light Scattering in Solids II, ed. by M. Cardona and G. Güntherodt*, Springer, Berlin, Heidelberg, New York, 1982.
16. W. Richter, *Springer Tracts in Modern Physics **Vol. 78**, Resonant Raman Scattering in Semiconductors, ed. by G. Höhler*, Springer, Berlin, Heidelberg, New York, 1976.
17. D. E. Aspnes, *Handbook of Semiconductors **Vol.2**, Ed. M. Balkanski*, North Holland Publishing Company, 1980.
18. M. Born and K. Huang, *Dynamical Theory of Crystal Lattices*, Clarendon Press, Oxford, 1954.
19. R. Loudon, *Proc. Royal Soc.* **A275**, 218 (1963).
20. A. Pinczuk and E. Burstein, *Topics in Applied Physics **Vol. 8**, Light Scattering in Solids, ed. by M. Cardona and G. Güntherodt*, Springer, Berlin, Heidelberg, New York, 1975.
21. A. Cantarero, C. Trallero-Giner, and M. Cardona, *Phys. Rev.* **B 39**, 8388 (1989).
22. P. Brüesch, *Springer Series in Solid-State Sciences **Vol. 34**, Phonons: Theory and Experiments I, ed. by M. Cardona, P. Fulde, H.J. Queisser*, Springer, Berlin, Heidelberg, New York, 1982.
23. H. Fröhlich, *Adv. Phys.* **3**, 325 (1954).
24. F. Stern, *Phys. Rev. Lett.* **18**, 546 (1967).
25. R. Trommer and M. Cardona, *Phys. Rev.* **B 17**, 1865 (1978).
26. M. Sinyukov, R. Trommer, and M. Cardona, *phys. stat. sol. (b)* **86**, 563 (1978).
27. A. Compaan and H. J. Trodahl, *Phys. Rev.* **B 29**, 793 (1984).
28. C. M. Brandmüller, *Springer Tracts in Modern Physics, Vol. 75, Ed. G. Höhler*, Springer Verlag Berlin, Heidelberg, New York, 1975.

29. W. Kress and e. F. W. de Wette, *Surface Phonons, Springer Series in Surface Sciences 21*, Sringer-Verlag, Berlin, Heidelberg, 1991.
30. R. Fuchs and K. L. Kliever, *Phys. Rev.* **140**, A2076 (1965).
31. L. Rayleigh, *Proc. London Math. Soc.* **17**, 4 (1887).
32. D. Farias and K. Rieder, *Rep. Prog. Phys.* **61**, 1575 (1998).
33. H. Ibach and D. L. Mills, *Electron Energy Loss Spectroscopy and Surface Vibrations*, Academic, New York, 1982.
34. K. Hinrichs, A. Schierhorn, P. Haier, N. Esser, W. Richter, and J. Sahm, *Phys. Rev. Lett.* **79**, 1094 (1997).
35. P. H. N. Esser, M. Köpp and W. Richter, *J. Electron Spectrosc. Relat. Phenom.* **64/65**, 85 (1993).
36. P. H. N. Esser, M. Köpp and W. Richter, *phys. stat. sol. (b)* **152**, 191 (1995).
37. M. A. Hines, T. D. Harris, A. L. Harris, and Y. C. Chabal, *J. Electron. Spectrosc. Rel. Phenom.* **64/65**, 183 (1993).
38. S. Ryjukov, T. Nagao, V. Lifshits, and S. Hasegawa, *Surface Science* **488**, 15 (1999).
39. H. W. Yeom, S. Takeda, E. Rotenberg, I. Matsuda, K. Horikoshi, J. Schaefer, C. M. Lee, S. D. Kevan, T. Ohta, T. Nagao, and S. Hasegawa, *Phys. Rev. Lett.* **82**, 4898 (1999).
40. S. Wang, W. Lu, W. G. Schmidt, and J. Bernholc, *Phys. Rev.* **B 68**, 035329 (2003).
41. J. Cho, D. Oh, K. Kim, and L. Kleinman, *Phys. Rev.* **B 64**, 235302 (2001).
42. O. Bunk, G. Falkenberg, J. H. Zeysing, L. Lottermoser, R. L. Johnson, M. Nielsen, F. Berg-Rasmussen, J. Baker, and R. Feidenhans'l, *Phys. Rev.* **B 59**, 12228 (1999).
43. F. Bechstedt, A. Krivosheeva, J. Furthmüller, and A. Stekolnikov, *Phys. Rev. B* **68**, 193406 (2003).
44. K. Sakamoto, H. Ashima, H. W. Yeom, and W. Uchida, *Phys. Rev.* **B 62**, 9923 (2000).
45. J. F. Geisz, D. J. Friedman, J. M. Olson, S. R. Kurtz, and B. M. Keyes, *Cryst. Growth* **195**, 401 (1998).
46. M. Kondow, S. Nakatsuka, T. Kitatani, Y. Yazawa, and M. Okai, *Jpn. J. Appl. Phys.* **35**, 5711 (1996).
47. J. S. Wang, A. R. Kovsh, L. Wei, J. Y. Chi, Y. T. Wu, P. Y. Wang, and V. M. Ustinov, *Nanotechnology* **12**, 430 (2001).
48. M. Weyers and M. Sato, *Appl. Phys. Lett.* **62**, 1396 (1993).
49. H. C. Alt, A. Y. Egorov, H. Riechert, B. Wiedemann, J. D. Meyer, R. W. Michelmann, and K. Bethge, *Appl. Phys. Lett.* **77**, 3331 (2000).
50. A. M. Mintairov, P. A. Balagnov, V. G. Melehin, N. N. Faleev, J. L. Merz, Y. Qiu, S. A. Nikishin, and H. Temkin, *Phys. Rev. B* **56**, 15863 (1997).
51. T. Prokofyeva, T. Sauncy, M. Seon, M. Holtz, Y. Qui, S. Nikishin, and H. Temkin, *Appl. Phys. Lett.* **73**, 1409 (1998).
52. B. H. Armstrong, *J. Quant. Spectrosc. Radiat. Transfer.* **7**, 61 (1967).
53. M. Herman, W. Richter, and H. Sitter, *Epitaxy, Physical Principles and Technical Implementation*, Springer-Verlag Berlin Heidelberg New York, 2004.
54. H. Richter, Z. Wang, and L. Ley, *Solid State Communications* **39**, 625 (1981).
55. R. Singh and S. Prakash, *Surface Science* **532**, 780 (2003).

56. S. Piscanec, A. Ferrari, M. Cantoro, S. Hofmann, J. Zapien, Y. Lifshitz, S. Lee, and J. Robertson, *Materials Science and Engineering* **C 23**, 931 (2003).
57. S. Bhattacharyya and S. Samui, *Appl. Phys. Lett.* **84**, 1564 (2004).
58. Y. Yang, X. Wu, G. Huang, D. Hu, and G. Siu, *Physics Letters* **A 338**, 379 (2005).
59. S. Guha, J. Menendez, J. B. Page, and G. B. Adams, *Phys. Rev.* **53**, 13106 (1996).
60. A. Zi, H. Büscher, C. Falter, W. Ludwig, K. Zhang, and X. Xie, *Appl. Phys. Lett.* **69**, 200 (1996).
61. T. Thonhauser and G. D. Mahan, *Phys. Rev.* **69**, 075213 (2004).
62. W. Cheng and S. Ren, *Phys. Rev.* **65**, 205305 (2002).
63. H. Tung and I. P. Herman, *Phys. Rev.* **B 43**, 2299 (1991).
64. A. C. Ferrari, S. Piscanec, S. Hoffmann, M. Cantoro, C. Ducati, and J. Robertson, *AIP Conf. Proc.* **685**, 507 (2003).
65. R. J. Nemanich, D. K. Biegelsen, R. A. Street, and L. E. Fennell, *Phys. Rev.* **B 29**, 6005 (1984).
66. J. Menendez and M. Cardona, *Phys. Rev.* **B 29**, 2051 (1984).
67. F. Cerdeira, E. Anastassakis, W. Kauschke, and M. Cardona, *Phys. Rev. Lett.* **57**, 3209 (1986).
68. F. Agullo-Rueda, E. E. Mendez, and J. M. Hong, *Phys. Rev.* **B 38**, 12720 (1988).
69. F. Caqlle, J. M. Calleja, F. Meseguer, C. Tejedor, L. Vina, and C. Lopez, *Phys. Rev.* **B 38**, 12720 (1988).
70. T. Seiyama, A. Kato, K. Fujiishi, and M. Nagatani, *Anal. Chem.* **34**, 102 (1962).
71. H. Z. Zhang, Y. C. Kong, Y. Z. Wang, X. Du, Z. G. Bai, J. J. Wang, D. P. Yu, Y. Ding, Q. L. Hang, and S. Q. Feng, *Solid State Communications* **109**, 677 (1962).
72. A. Hartschuh, E. Sanchez, X. Xie, and L. Novotny, *Phys. Rev. Lett.* **90**, 095503 (2003).
73. J. F. Scott, *Phys. Rev. B* **2**, 1209 (1970).
74. R. H. Callender, S. S. Sussman, M. Selders, and R. K. Chang, *Phys. Rev. B* **7**, 3788 (1973).
75. J. M. Calleja and M. Cardona, *Phys. Rev. B* **16**, 3753 (1977).
76. C. A. Arguello, D. L. Rousseau, and S. P. S. Porto, *Phys. Rev.* **181**, 1351 (1969).
77. M. Rajalakshmi, A. K. Arora, B. S. Bendre, and S. Mahamuni, *J. Appl. Phys.* **87**, 2445 (2000).
78. F. Pedreschi, J. D. OMahonya, P. Weightman, and J. R. Power, *Appl. Phys. Lett.* **73**, 2152 (1998).
79. K. Fleischer, S. Chandola, N. Esser, W. Richter, J. McGilp, W. Schmidt, S. Wang, W. Lu, and J. Bernholc, *Applied Surface Science* **234**, 302 (2004).

CALCULATION OF REFLECTANCE ANISOTROPY FOR SEMICONDUCTOR SURFACE EXPLORATION

W. G. SCHMIDT

Friedrich-Schiller-Universität,
Institut für Festkörpertheorie und -optik,
Max-Wien-Platz 1, 07743 Jena, Germany
E-mail: W.G.Schmidt@ifto.physik.uni-jena-de

Reflectance Anisotropy Spectroscopy (RAS) is exquisitely sensitive to probe surfaces, with many potential applications in determining surface geometries or monitoring material growth. Thanks to recent computational and methodological progress it has now become possible to calculate surface optical spectra accurately and with true predictive power. Here I review briefly the simulation of RAS spectra for semiconductor surfaces and discuss recent methodological advances, which allow for the modeling of self-energy, excitonic and local-field effects in large and complex systems.

1. Theory of Reflectance Anisotropy

Following earlier pioneering works[1,2] Del Sole[3] obtained an expression for the surface contribution to the reflectance, $\Delta R/R$, where R is the reflectance according to the Fresnel equation. For s-light polarized along i and normal incidence it holds

$$\frac{\Delta R_i}{R}(\omega) = \frac{4\omega}{c}\Im\left\{\frac{\Delta\epsilon_{ii}(\omega)}{\epsilon_b(\omega)-1}\right\}, \tag{1}$$

where ϵ_b is the bulk dielectric function, and

$$\begin{aligned}\Delta\epsilon_{ij} = &\int dzdz' \left[\epsilon_{ij}(\omega;z,z') - \delta_{ij}\delta(z-z')\epsilon_0(\omega;z)\right] \\ &- \int dzdz'dz''dz''' \epsilon_{iz}(\omega;z,z')\epsilon_{zz}^{-1}(\omega;z',z'')\epsilon_{zj}(\omega;z'',z''').\end{aligned} \tag{2}$$

Here $\epsilon_{ij}(\omega;z,z')$ is the non-local *macroscopic* dielectric tensor of the solid-vacuum interface accounting for all many-body and local-field effects[4]. Eq. (2) can be evaluated by replacing the semi-infinite crystal by an artificial super-cell, large enough to represent the vacuum as well as the surface and bulk regions of the crystal under investigation. Provided that: (i) the slab is large enough to properly describe the surface region of the crystal, i.e., the surface as well as surface-modified bulk wavefunctions and (ii) the off-diagonal terms of the dielectric tensor are small compared to the diagonal ones, a simple expression for the surface contribution to the reflectivity can be derived[5]

$$\frac{\Delta R_i}{R}(\omega) = \frac{4\omega}{c}\mathrm{Im}\left\{\frac{4\pi\alpha_{ii}^{hs}(\omega)}{\epsilon_b(\omega)-1}\right\}. \tag{3}$$

Here $\alpha_{ii}^{hs}(\omega)$ with $i = x, y$ is the diagonal tensor component of the averaged half-slab polarizability. Under the conditions mentioned above, Eq. (3) contains in principle *all* surface contributions to the optical reflectance. This includes contributions due to surface electronic states, atomic relaxations, the influence of the surface potential on the bulk wavefunctions as well as surface local-field effects and many-body effects such as the electronic self-energy and the electron-hole attraction. Of course, it is not an easy task to include all these contributions in practical calculations of the slab polarizability.

Before discussing the numerical calculation of α it is in order to mention one point concerning the comparison with experimental data. Optical reflectance or reflectivity is defined as the square of the modulus of the reflection amplitude, $R = |r|^2$, i.e., it gives the ratio of the reflected to incident beam intensity. Care is required here, because the use of R and r is sometimes reversed. Most experimentalists publish spectra of the real part of the reflection amplitude, $\Re(\Delta r/r)$. These are sometimes directly compared with calculations for $\Delta R/R$. This leads to seemingly overestimated anisotropies, since

$$\begin{aligned}\Re\left\{\frac{\Delta r}{r}\right\} &= 2\Re\left\{\frac{r_x - r_y}{r_x + r_y}\right\} = 2\frac{|r_x|^2 - |r_y|^2}{|r_x|^2 + |r_y|^2 + 2\Re\left\{r_x r_y^*\right\}} \\ &\approx \frac{|r_x|^2 - |r_y|^2}{|r_x|^2 + |r_y|^2} = \frac{R_x - R_y}{R_x + R_y} = \frac{1}{2}\frac{\Delta R}{R}.\end{aligned} \tag{4}$$

Here I used $r_x r_y^* = |r|^2 - \frac{1}{4}|\Delta r|^2 \approx |r|^2 = R$.

2. Computational

The state-of-the-art approach to the microscopic calculation of surface optical properties consists of three steps. (i) The structurally relaxed ground state of the surface is calculated, most often within density functional theory (DFT) employing either the local density approximation (LDA) or the generalized gradient approximation (GGA) for the description of the electron exchange and correlation interaction. These calculations yield the eigenenergies and eigenfunctions used to represent the one- and two-body Green's functions. (ii) The electronic self-energy is obtained within the GW approximation (GWA), which takes the discontinuity of the electron-electron exchange and correlation energy on addition of an electron into account, and allows for calculating accurate single-quasiparticle spectra. (iii) The screened electron-hole attraction and the unscreened electron-hole exchange are taken into account by solving the Bethe-Salpeter equation (BSE) for the macroscopic polarization, shown schematically in Fig. 1. From the BSE we obtain the linear optical spectrum including the effects of many-body electron-electron and electron-hole interaction. No assumption about the atomic configuration or the electronic properties of the system needs to be made; they are calculated self-consistently and without any input from experiment. In the following I will sketch the specifics of the numerical implementation which has been developed in Jena[6].

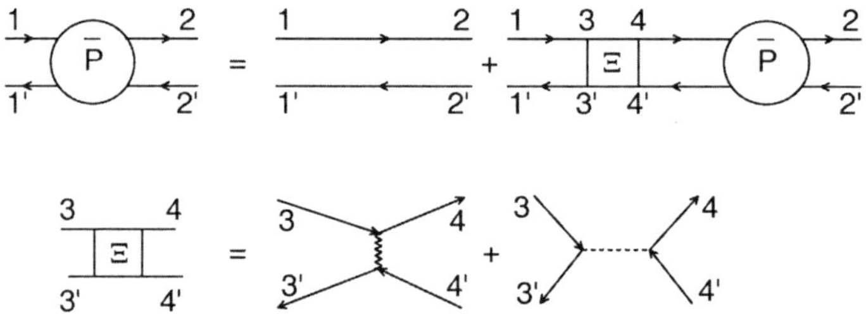

Figure 1. Diagrammatic representation of the Bethe-Salpeter equation. The macroscopic polarization is related to the polarization of independent particles (lines $1 \rightarrow 2$ and $1 \rightarrow 2$) and an interaction kernel Ξ containing the screened electron-hole attraction (wriggled line) and the unscreened electron-hole exchange (dashed line).

We start from *first-principles* pseudopotential calculations, using a massively parallel real-space finite-difference DFT implementation[7]. A multigrid technique is used for convergence acceleration. Electronic self-energy effects are included by postprocessing the DFT results. The local exchange and correlation potential $V^{XC}(\mathbf{r})$ is formally replaced by the nonlocal and energy-dependent self-energy operator $\Sigma(\mathbf{r}, \mathbf{r}'; E)$[8]. For the calculation of Σ we use the GW approximation[9,10], where the self-energy operator is expressed as convolution $\Sigma = iGW$ of the dynamically screened Coulomb potential W and the single-particle propagator G. The calculation of surface optical spectra usually involves a very large number of electronic states. Therefore we introduce further approximations[11,12]: the GW quasiparticle energies are obtained from the DFT eigenvalues in a perturbative manner by

$$\varepsilon_n(\mathbf{k})^{QP} = \varepsilon_n(\mathbf{k}) + \frac{1}{1+\beta_{n,\mathbf{k}}}\left[\Sigma^{st}_{n,\mathbf{k}} + \Sigma^{dyn}_{n,\mathbf{k}}\left(\varepsilon_n(\mathbf{k})\right) - V^{XC}_{n,\mathbf{k}}\right], \tag{5}$$

where the self-energy operator Σ has been divided into static (st) and dynamic (dyn) contributions. Indices at Σ and V^{XC} indicate diagonal matrix elements with the respective wave functions. $\beta_{n,\mathbf{k}}$ is the linear coefficient in the expansion of Σ^{dyn} around the DFT eigenvalue $\varepsilon_n(\mathbf{k})$. The static part is split into two parts

$$\Sigma^{st}(\mathbf{r},\mathbf{r}') = \frac{1}{2}\sum_{n,\mathbf{k}} \psi_{n,\mathbf{k}}(\mathbf{r})\psi^*_{n,\mathbf{k}}(\mathbf{r}')\left[W(\mathbf{r},\mathbf{r}';0) - v(\mathbf{r}-\mathbf{r}')\right] - \sum_{v,\mathbf{k}} \psi_{v,\mathbf{k}}(\mathbf{r})\psi^*_{v,\mathbf{k}}(\mathbf{r}')W(\mathbf{r},\mathbf{r}';0)\ , \tag{6}$$

representing the Coulomb hole Σ^{COH} and the screened exchange Σ^{SEX}. The $\psi_{n,\mathbf{k}}$ are the DFT-LDA wave functions. Σ^{SEX} contains a sum over the occupied valence states v only. The major bottleneck in the GW calculation is the computation of the screened interaction W. An extreme speedup can be achieved by using a model dielectric function. We use the version suggested by Bechstedt *et al.*[12]

$$\epsilon(\mathbf{q},\rho) = 1 + \left[(\epsilon_\infty - 1)^{-1} + \left(\frac{q}{q_{TF}(\rho)}\right)^2 + \frac{3q^4}{4k_F^2(\rho)q_{TF}^2(\rho)}\right]^{-1}, \tag{7}$$

where k_F and q_{TF} represent the Fermi and Thomas-Fermi wave-vectors, respectively, which depend on the electron density ρ. Together with the LDA-like ansatz of Hybertsen and Louie[11] for approximating the spatial dependence of the screening of the inhomogeneous system

$$W(\mathbf{r},\mathbf{r}';0) = \frac{1}{2}\left[W^h\left(\mathbf{r}-\mathbf{r}',\rho(\mathbf{r})\right) + W^h\left(\mathbf{r}-\mathbf{r}',\rho(\mathbf{r}')\right)\right] \tag{8}$$

by that of a homogeneous electron gas W^h, Eq. (7) allows for an analytic solution for Σ_{COH}. The static Coulomb hole contribution to the self-energy takes the form of a local potential

$$\Sigma^{COH}(\mathbf{r}) = -\frac{q_{TF}(\mathbf{r})}{2}\sqrt{1-\frac{1}{\epsilon_\infty}}\left[1 + \frac{q_{TF}(\mathbf{r})}{k_F(\mathbf{r})}\sqrt{\frac{3\epsilon_\infty}{\epsilon_\infty - 1}}\right]^{-\frac{1}{2}}, \tag{9}$$

where k_F and q_{TF} are computed at the local density $\rho(\mathbf{r})$.

The matrix elements $\Sigma^{SEX}_{n,\mathbf{k}}$ are calculated in reciprocal space. However, only the diagonal elements in the Fourier transform of W are retained and its nonlocality is approximated by using state-averaged electron densities

$$\rho_{n,\mathbf{k}} = \int d\mathbf{r}^3 \rho(\mathbf{r}) |\psi_{n,\mathbf{k}}(\mathbf{r})|^2 , \tag{10}$$

in the calculation of k_F and q_{TF}[12]. Finally, the dynamic terms $\beta_{n,\mathbf{k}}$ and Σ^{dyn} in (5), are approximated by simple integrals of the dielectric function[12]. For the actual calculations we use (7) together with a single-plasmon-pole approximation to describe the frequency dependence. Local-field effects are again included using the mean-density approximation (10). The integrals are numerically evaluated for a dense sampling of ρ and the results for $\beta_{n,\mathbf{k}}(\rho)$ and $\Sigma^{dyn}(\rho)$ are fitted to polynomials. These are then used for a fast computation of the dynamic contributions to the self-energy during the actual GW calculations. For several III-V compounds and their surfaces this approximate treatment of self-energy corrections has been shown to result in excitation energies which are within about 0.1 eV of the experimental values[13,14,15].

Excitation energies obtained within the quasiparticle formalism describe one-particle excitations, such as those involved in direct or inverse photoemission experiments. For the description of the optical absorption process, however, one needs to go beyond this single-quasiparticle level. The polarization function P including electron-hole attraction and local-field effects (or electron-hole exchange) can be obtained from the solution of the Bethe-Salpeter equation[16,17,18,19]),

$$\bar{P} = P_0 + P_0(\bar{v} - W)rP , \tag{11}$$

where $\bar{v}$ is the bare Coulomb potential without its long-range part and P_0 represents the polarization function in random-phase, or, more precisely, independent-quasiparticle approximation. The macroscopic polarizability is obtained from the Fourier transform of the diagonal part of P.

A convenient and natural basis for solving Eq. (11) is given by the orthonormal and complete set of Bloch functions defined by the Kohn-Sham problem. If P_0 is explicitly expressed in terms of Bloch functions and quasiparticle energies and transformed into Bloch space, the solution of the BSE (11) can be written in resolvent representation as

$$\bar{P}_{(n_1,n_2)(n_3,n_4)} = \left[\hat{H} - \omega\right]^{-1}_{(n_1,n_2)(n_3,n_4)} (f_{n_4} - f_{n_3}), \tag{12}$$

where the two-particle Hamiltonian

$$\hat{H}_{(n_1,n_2)(n_3,n_4)} \equiv (\varepsilon_{n_1}^{QP} - \varepsilon_{n_2}^{QP})\delta_{(n_1,n_3)}\delta_{(n_2,n_4)} + (f_{n_2} - f_{n_1})\times$$
$$\int d\mathbf{r}_1 d\mathbf{r}_2 d\mathbf{r}_3 d\mathbf{r}_4 \psi_{n_1}(\mathbf{r}_1)\psi_{n_2}^*(\mathbf{r}_2)\psi_{n_3}^*(\mathbf{r}_3)\psi_{n_4}(\mathbf{r}_4) \times$$
$$[\delta(\mathbf{r}_1 - \mathbf{r}_2)\delta(\mathbf{r}_3 - \mathbf{r}_4)\bar{v}(\mathbf{r}_1 - \mathbf{r}_3) - \delta(\mathbf{r}_1 - \mathbf{r}_3)\delta(\mathbf{r}_2 - \mathbf{r}_4)W(\mathbf{r}_1, \mathbf{r}_2)] \tag{13}$$

has been introduced. The $f_n = 0, 1$ is the occupation number of the state n, denoting both band index and wave vector. The dimension of $\hat{H}$ can be reduced by a factor of two if one observes that due to the factors $(f_{n_4} - f_{n_3})$ in (12) and $(f_{n_2} - f_{n_1})$ in Eq. (13), only pairs containing one filled and one empty Bloch state contribute to the macroscopic polarization. A further reduction of the dimension by a factor of two can be achieved when the off-diagonal blocks coupling the hermitian resonant part of $\hat{H}$,

$$\hat{H}^{res}_{v c\mathbf{k},v'c'\mathbf{k}'} = (\varepsilon_{c\mathbf{k}}^{QP} - \varepsilon_{v\mathbf{k}}^{QP})\delta_{vv'}\delta_{cc'}\delta_{\mathbf{k},\mathbf{k}'} +$$
$$2\int d\mathbf{r}_1 d\mathbf{r}_2 \psi_{c\mathbf{k}}^*(\mathbf{r}_1)\psi_{v\mathbf{k}}(\mathbf{r}_1)\bar{v}(\mathbf{r}_1 - \mathbf{r}_2)\psi_{c'\mathbf{k}'}(\mathbf{r}_2)\psi_{v'\mathbf{k}'}^*(\mathbf{r}_2) -$$
$$\int d\mathbf{r}_1 d\mathbf{r}_2 \psi_{c\mathbf{k}}^*(\mathbf{r}_1)\psi_{c'\mathbf{k}'}(\mathbf{r}_1)W(\mathbf{r}_1, \mathbf{r}_2)\psi_{v\mathbf{k}}(\mathbf{r}_2)\psi_{v'\mathbf{k}'}^*(\mathbf{r}_2), \tag{14}$$

and the anti-resonant part, $-\left[\hat{H}^{res}\right]^*$, are neglected. Apart from special cases, e.g. the calculation of plasmon resonances where the mixing of inter-band transitions of both positive and negative frequencies must be included in the calculations[20], the coupling can be neglected in the calculation of optical properties[21,22]. Additional approximations in (14) are the restriction to spin-singlets, static screening and direct transitions, i.e., the neglect of momentum transfer by photons. If furthermore umklapp processes are neglected, the exciton Hamiltonian can be calculated in reciprocal space according to

$$\hat{H}^{res}_{v c\mathbf{k},v'c'\mathbf{k}'} = (\varepsilon_{c\mathbf{k}}^{QP} - \varepsilon_{v\mathbf{k}}^{QP})\delta_{vv'}\delta_{cc'}\delta_{\mathbf{k}\mathbf{k}'} \tag{15}$$
$$+\frac{4\pi}{\Omega}\sum_{\mathbf{G},\mathbf{G}'}\{\,2\frac{\delta_{\mathbf{G}\mathbf{G}'}(1-\delta_{\mathbf{G}\mathbf{0}})}{|\mathbf{G}|^2}B_{cv}^{\mathbf{k}\mathbf{k}}(\mathbf{G})B_{c'v'}^{\mathbf{k}'\mathbf{k}'*}(\mathbf{G}) -$$
$$\frac{\epsilon^{-1}(\mathbf{k}-\mathbf{k}'+\mathbf{G},\mathbf{k}-\mathbf{k}'+\mathbf{G}',0)}{|\mathbf{k}-\mathbf{k}'+\mathbf{G}|^2}B_{cc'}^{\mathbf{k}\mathbf{k}'}(\mathbf{G})B_{vv'}^{\mathbf{k}\mathbf{k}'*}(\mathbf{G}')\,\}, \tag{16}$$

where the Bloch integral

$$B_{nn'}^{\mathbf{kk}'}(\mathbf{G}) = \frac{1}{\Omega}\int d\mathbf{r} u_{n\mathbf{k}}^{*}(\mathbf{r}) e^{i\mathbf{Gr}} u_{n'\mathbf{k}'}(\mathbf{r}) \tag{17}$$

over the periodic parts u of the Bloch wave functions has been introduced and Ω denotes the volume of the unit cell. We replace the inverse dielectric matrix ϵ^{-1} by the same diagonal model dielectric function due to Bechstedt (7), which has been used in the calculation of the self-energy operator. The influence of the off-diagonal elements is again approximated by using state-dependent electron densities in (7), which were calculated using the mean-density approximation (10).

After the Hamiltonian has been calculated, one needs to determine the frequency-dependent polarizability. The usual approach consists in transforming the resolvent into an effective eigenvalue problem, which is then solved by diagonalization [21,23]. In detail, using the spectral representation

$$\left[\hat{H} - \omega\right]^{-1} = \sum_{\lambda,\lambda'} \frac{|A^{\lambda}\rangle S_{\lambda,\lambda'}^{-1} \langle A^{\lambda'}|}{E_{\lambda} - w}, \tag{18}$$

where $|A^{\lambda}\rangle$ and E_{λ} are the eigenvectors and eigenvalues of the exciton Hamiltonian

$$\hat{H}|A^{\lambda}\rangle = E_{\lambda}|A^{\lambda}\rangle \quad \text{and} \quad S_{\lambda,\lambda'} = \langle A^{\lambda'}|A^{\lambda}\rangle, \tag{19}$$

the diagonal components of the macroscopic polarizability are given by

$$\alpha_{jj}^{M}(\omega) = \frac{4e^2\hbar^2}{\Omega} \sum_{\lambda} \left| \sum_{\mathbf{k}} \sum_{c,v} \frac{\langle c\mathbf{k}|v_j|v\mathbf{k}\rangle}{\varepsilon_c(\mathbf{k}) - \varepsilon_v(\mathbf{k})} A_{vc\mathbf{k}}^{\lambda} \right|^2 \times \left\{ \frac{1}{E_{\lambda} - \hbar(\omega + i\gamma)} + \frac{1}{E_{\lambda} + \hbar(\omega + i\gamma)} \right\}, \tag{20}$$

where v_j is the corresponding Cartesian component of the single-particle velocity operator and γ the damping constant. Here, the contributions of the anti-resonant part of the exciton Hamiltonian have been formally included, while the coupling parts are neglected.

The dimension of the exciton Hamiltonian $N = N_v \cdot N_c \cdot N_{\mathbf{k}}$ is about $10^5 ... 10^6$ already for the relatively small unit cell of an unreconstructed surface. Even with today's powerful supercomputers, the diagonalization of matrices of this size, which scales as $\mathcal{O}(N^3)$, is prohibitively slow.

Therefore, we formulate the calculation of the ω-dependent polarizability as an initial-value problem. If a vector $|\mu^j\rangle$ of dipole moments with elements

$$\mu^j_{v c \mathbf{k}} = \frac{\langle c\mathbf{k}|v_j|v\mathbf{k}\rangle}{\varepsilon_c(\mathbf{k}) - \varepsilon_v(\mathbf{k})} \tag{21}$$

is introduced, Eq. (20) takes the form

$$\alpha^M_{jj}(\omega) = \frac{4e^2\hbar^2}{\Omega}\sum_\lambda \left|\langle\mu^j|A^\lambda\rangle\right|^2 \left\{\frac{1}{E_\lambda - \hbar(\omega+i\gamma)} + \frac{1}{E_\lambda + \hbar(\omega+i\gamma)}\right\}. \tag{22}$$

This is equivalent to the Fourier representation

$$\alpha^M_{jj}(\omega) = \frac{4e^2\hbar^2}{\Omega} i \int_0^\infty dt e^{i(\omega+i\gamma)t}\left\{\langle\mu^j|\xi^j(t)\rangle - \langle\mu^j|\xi^j(t)\rangle^*\right\}, \tag{23}$$

where the time evolution of the vector $|\xi^j(t)\rangle$ is driven by the pair Hamiltonian

$$i\hbar\frac{d}{dt}|\xi^j(t)\rangle = \hat{H}|\xi^j(t)\rangle \tag{24}$$

and the initial vector elements are given by

$$|\xi^j(0)\rangle = |\mu^j\rangle. \tag{25}$$

The equivalence[24] can be shown by integrating $|\xi(t)\rangle = e^{\frac{\hat{H}t}{i\hbar}}|\mu\rangle$ and exploiting the spectral representation as in (18). We solve the initial-value problem defined by (24) and (25) using the central-difference method which obtains $|\xi(t_{i+2})\rangle$ from $|\xi(t_i)\rangle$ and $|\xi(t_{i+1})\rangle$ by an explicit scheme

$$\hat{H}|\xi(t_{i+1})\rangle = i\hbar\frac{|\xi(t_{i+2})\rangle - |\xi(t_i)\rangle}{2\Delta t}. \tag{26}$$

This procedure only requires one matrix-vector multiplication per time step. The stability of the difference scheme (26) requires that $\Delta t < \hbar/||\hat{H}||$. The upper limit of the Fourier integral (23) can be truncated, due to the exponential $e^{-\gamma t}$. Therefore, the number of time steps, i.e., matrix-vector multiplications, is nearly independent of the dimension of the system and governed by γ. Of the order of 10^3 time steps are typically required using a broadening parameter $\gamma = 0.1$ eV. The operation count for this method scales thus as $\mathcal{O}(N^2)$, compared to the $\mathcal{O}(N^3)$ for the matrix diagonalization. The cross-over point for the CPU time usage of both methods in our implementation is reached for a number of electron-hole pair states as low as about 2000. Moreover, the matrix-vector multiplications can be easily distributed on several processors of a parallel computer. Hence it is possible to include all relevant many-body effects from *first-principles* in numerical calculations of the surface optical response[25,26].

3. Example: Si(110) surface

The influence of many-body effects on the optical anisotropy is demonstrated for the hydrogen-passivated Si(110) surface. It is one of the first systems studied by RAS[27] and has become a calibration standard for RAS apparatus and a textbook example for surface optical properties[28].

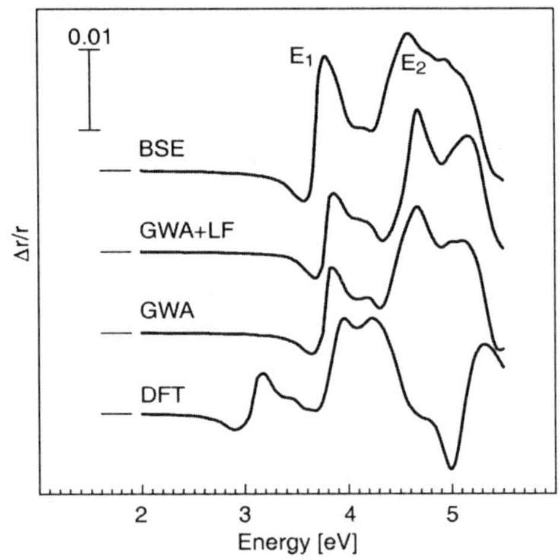

Figure 2. RAS $[Re\{(r_{[1\bar{1}0]} - r_{[001]})/ < r >\}]$ calculated for Si(110):H within DFT-LDA, in GWA, in GWA with the effects of local fields included, and from the solution of the BSE.

Fig. 2 contains the calculated RAS spectra[25] for the Si(110):H surface represented by a 12-layer slab. The DFT-LDA spectrum shows two strong positive RAS features near the E_1 and E_2 bulk critical point energies. However, the features are far too broad. Inclusion of quasiparticle effects in GW approximation leads to a blue-shift of the spectrum by about 0.6 - 0.7 eV and changes the line shape. In particular the anisotropy at the E_1 energy is enhanced compared to the DFT-LDA spectrum. The E_1 peak height relative to the E_2 anisotropy is still much smaller than measured, however. LF effects lead to surprisingly small changes of the spectrum. A drastic enhancement of the optical anisotropy at the E_1 energy and a red-shift of the entire spectrum by about 0.1 – 0.2 eV result, however, from the inclusion of the attractive electron-hole interaction. This is shown by the uppermost spectrum in Fig. 2. Also the characteristic negative anisotropy below the E_1 energy is enhanced by excitonic effects.

Although a substantial and systematic improvement of the calculated spectrum occurs upon inclusion of many-body effects, no complete reproduction of the experiment[27] is possible. From numerical tests on the single-particle level of theory[29], the remaining discrepancies can be traced to the insufficient **k**-point sampling and a slab which is too thin to allow for the complete description of the surface-perturbed bulk wave functions responsible for the observed optical anisotropies.

Acknowledgement

I thank Friedhelm Bechstedt, Jerry Bernholc, and Patrick H. Hahn for many useful discussions.

References

1. J. D. E. McIntyre and D. E. Aspnes, Surf. Sci. **24**, 417 (1971).
2. A. Bagchi, R. G. Barrera, and A. K. Rajagopal, Phys. Rev. B **20**, 4824 (1979).
3. R. Del Sole, Solid State Commun. **37**, 537 (1981).
4. R. Del Sole and E. Fiorino, Phys. Rev. B **29**, 4631 (1984).
5. F. Manghi, R. Del Sole, A. Selloni, and E. Molinari, Phys. Rev. B **41**, 9935 (1990).
6. for download see software link at http://www.ifto.uni-jena.de/en/begin.html.
7. J. Bernholc *et al.*, phys. stat. sol. (b) **217**, 685 (2000).
8. F. Aryasetiawan and O. Gunnarsson, Rep. Prog. Phys. **61**, 237 (1998).
9. L. Hedin, Phys. Rev. **139**, A769 (1965).
10. L. Hedin and S. Lundqvist, in *Solid State Physics*, edited by H. Ehrenreich, F. Seitz, and D. Turnball (Academic, New York, 1969), Vol. 23, p. 1.
11. M. S. Hybertsen and S. G. Louie, Phys. Rev. B **37**, 2733 (1988).
12. F. Bechstedt, R. Del Sole, G. Cappellini, and L. Reining, Solid State Commun. **84**, 765 (1992).
13. W. G. Schmidt, J. L. Fattebert, J. Bernholc, and F. Bechstedt, Surf. Rev. Lett. **6**, 1159 (1999).
14. W. G. Schmidt *et al.*, Phys. Rev. B. **61**, R16335 (2000).
15. W. G. Schmidt *et al.*, phys. stat. sol. **188**, 1401 (2001).
16. L. J. Sham and T. M. Rice, Phys. Rev. **144**, 708 (1966).
17. W. Hanke and L. J. Sham, Phys. Rev. B **21**, 4656 (1980).
18. G. Strinati, Rivista del Nuovo Cimento **11**, 1 (1988).
19. H. Stolz, *Einführung in die Vielelektronentheorie der Kristalle* (Akademie-Verlag, Berlin, 1974).
20. V. Olevano and L. Reining, Phys. Rev. Lett. **86**, 5962 (2001).
21. S. Albrecht, L. Reining, R. Del Sole, and G. Onida, Phys. Rev. Lett. **80**, 4510 (1998).
22. S. Albrecht, Ph.D. thesis, École Polytechnique, Paris, 1999.
23. M. Rohlfing and S. G. Louie, Phys. Rev. Lett. **80**, 3320 (1998).
24. S. Glutsch, D. S. Chemla, and F. Bechstedt, Phys. Rev. B **54**, 11592 (1996).
25. P. H. Hahn, W. G. Schmidt, and F. Bechstedt, Phys. Rev. Lett. **88**, 016402 (2002).
26. W. G. Schmidt, S. Glutsch, P. H. Hahn, and F. Bechstedt, Phys. Rev. B **67**, 085307 (2003).
27. D. E. Aspnes and A. A. Studna, Phys. Rev. Lett. **54**, 1956 (1985).
28. P. Y. Yu and M. Cardona, *Fundamentals of Semiconductors* (Springer-Verlag, Berlin, 1999).
29. W. G. Schmidt and J. Bernholc, Phys. Rev. B **61**, 7604 (2000).

MOLECULAR ASSEMBLY AT METAL SURFACES STUDIED BY REFLECTION ANISOTROPY SPECTROSCOPY

DAVID S. MARTIN

Department of Physics and Surface Science Research Centre,
University of Liverpool,
Liverpool, L69 3BX, England
E-mail: davidm@liverpool.ac.uk

Reflection anisotropy spectroscopy (RAS) is a non-destructive optical probe that can be used to investigate surfaces in a wide range of environments. RAS has been used to monitor the first stages of molecular assembly on metal surfaces in ultra-high vacuum and at the solid/liquid interface. Sensitivity to adsorbate-surface bonding and molecular orientation has been demonstrated. The Cu(110) and Au(110) surfaces have been the focus for RAS studies of molecular assembly and in this article, some aspects of the RA response of Cu(110) and Au(110) are described followed by a review of RAS studies of molecular assembly at these surfaces.

1. Introduction

Reflection anisotropy spectroscopy (RAS) is a non-destructive optical probe that can be used to investigate surfaces in ultra-high vacuum (UHV), ambient and liquid environments. RAS measures the difference in reflection of normal incidence plane-polarised light between two orthogonal directions in the surface plane [1,2]. When applied to cubic crystals, RAS achieves surface sensitivity since the optical response of the bulk crystal is isotropic by symmetry, and the signal arises from the lower symmetry of the surface. The interpretation of RA spectra of metal surfaces has benefited from experiments performed on clean surfaces in ultra-high vacuum (UHV) where results from techniques such as photoemission (PE) and inverse photoemission spectroscopy (IPES), low-energy electron diffraction (LEED) and scanning tunnelling microscopy (STM) are combined to reveal relationships between the RA response and the atomic and electronic structure and morphology of the surface.

RAS has been used to monitor molecular assembly at metal surfaces in UHV [3,4,5,6,7] and at the solid/liquid interface [8,9]. The Cu(110) and Au(110) surfaces have been the focus for RAS studies of molecular assembly in UHV and liquid environments, respectively. In this article, some aspects of the RA response of the clean surfaces of Cu(110) and Au(110) are summarised followed by a review of RAS studies of molecular assembly at these surfaces.

2. The (110) surface of FCC metals

The (110) surfaces of cubic materials have an intrinsic structural anisotropy and it is for this reason that (110) surfaces of metals have been the focus of RAS studies. The unit cell of the FCC structure and the associated (110) plane are shown in Fig. 1a. The atomic structure of the (110) surface shown in Fig. 1b represents a perfect termination of the bulk crystal structure along the (110) plane and is labelled a (1×1) structure. The clean Cu(110) surface exhibits the (1×1) surface structure. Surfaces sometimes reconstruct to a structure different from – yet related to – a perfect termination. One example of surface reconstruction is the 'missing-row' (1×2) structure (Fig. 1c) where every other $[1\bar{1}0]$ row of atoms from the (1×1) structure is missing and the length of the surface unit cell in the [001] direction is doubled. The clean Au(110) surface at room temperature adopts the (1×2) structure.

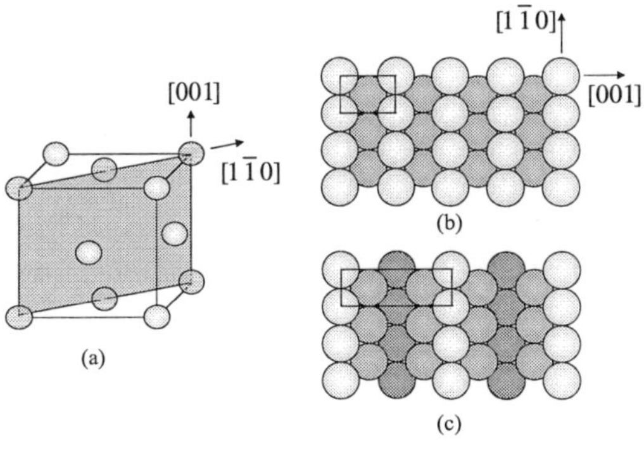

Figure 1. (a) Unit cell of the FCC structure and (110) plane (shaded). Perfect (110) surfaces of an FCC crystal: (b) (1×1) structure and (c) missing-row (1×2) structure.

The truncation of the periodic crystal structure at the surface introduces intrinsic surface states and modifies the character of the continuum states at the surface. RAS probes optical transitions that yield information on surface electronic structure and features in RA spectra of metal surfaces have been identified with transitions between surface states [10,11,12,13] and surface-modified bulk states [14,15]. This will become apparent when RAS of Cu(110) and Au(110) are discussed.

3. The RAS technique

The standard RA spectrometer was developed by Aspnes *et al* [1] and utilises the basic components of ellipsometry with the incorporation of a photoelastic modulator. RAS probes the difference in complex Fresnel reflection amplitudes r_x and r_y associated with two orthogonal in-plane surface directions x and y, respectively. The complex RA is defined as the difference in reflectance (Δr) normalised to the mean reflectance (r). For (110) surfaces, the two orthogonal in-plane directions $[1\bar{1}0]$ and $[001]$ (Fig. 1) have unequal reflection coefficients and are chosen as x and y. In this article the RA is defined as:

$$\frac{\Delta r}{r} = \frac{2(r_{[1\bar{1}0]} - r_{[001]})}{r_{[1\bar{1}0]} + r_{[001]}} \tag{1}$$

RAS is achieved by focusing plane-polarised light at normal incidence onto the surface of the specimen that is oriented such that its optical axes (x,y) are at 45° to the plane of light polarisation. The elliptically polarised reflected light is Fourier analysed to yield the real and imaginary parts of the RA. The propagation of light through the RA spectrometer and the analysis of the RA signal have been reviewed [2]. RA spectra are recorded sequentially, wavelength by wavelength, typically over the range 1.5 to 5.0 eV. Alternatively, transients can be monitored at a single wavelength on a timescale of ~20 ms. The real part of the RA is most widely reported from UHV studies due to the measurement of the imaginary part being sensitive to window strain [2].

4. RAS of the Cu(110) surface

The RA spectrum of clean Cu(110) at room temperature is shown in Fig. 2a. The spectrum is characterised by features at 2.1 eV, 4.25 eV and 4.9 eV (labelled A to C in Fig. 2a). The intense positive peak A arises from two different contributions: (i) a dominant contribution from transitions between surface states occurring at the $\bar{Y}$ symmetry point [10,13] and (ii) a smaller contribution assigned to local-field effects at the surface [13,16]. The surface electronic structure of Cu(110) in the region of the $\bar{Y}$ point of the surface Brillouin zone is shown in Fig. 2b [17]. In room temperature PE and IPE results, two surface states are found in the bulk band gap occurring in the region of the Fermi level (E_F) at $\bar{Y}$: an occupied surface state at binding energy ~0.4 eV below E_F [18,19] and an unoccupied surface state at ~1.8 eV above E_F [18,20,21]. These surface states, shown in Fig. 2b, have the correct difference in energy to account for the RAS peak A at 2.1 eV and appropriate symmetries; the occupied state has predominantly p_y character and the unoccupied state has predominantly *sp* character with respect to the Cu atoms in the surface layer. As a result, transitions at $\bar{Y}$ that generate peak A are induced only by light polarised along [001].

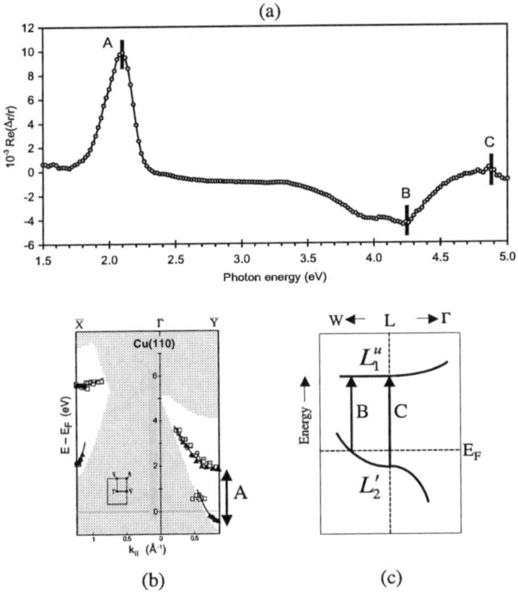

Figure 2. (a) RAS of clean Cu(110). (b) Surface electronic structure of Cu(110) showing data from [18] and [20], reprinted from [17], copyright 1998, with permission from Elsevier. Transitions between the two surface states (solid lines) in the region of $\bar{Y}$ give rise to peak A in (a). (c) Band structure at L indicating the transitions responsible for peaks B and C in (a).

Peaks B and C have been assigned [14] to transitions involving surface-modified bulk bands near E_F at the L symmetry point (Fig. 2c). Peak B originates from the transitions $E_F \rightarrow L_1^u$ and peak C involves the transitions

$L_2' \to L_1^u$. These assignments are supported by the temperature dependence of the energy position of RAS peaks B and C, which are found to agree with those of the transition energies between the associated bulk bands as determined from thermovariation optical spectroscopy [22]. An enhanced intensity of peak B has been associated with the presence of [001]-oriented surface steps [13] and the intensity and sign of peak B observed from vicinal Cu surfaces is influenced by the step density and coordination of corner atoms at step edges, respectively [23].

The RAS peak A provides a sensitive probe of the surface states associated with this peak. The process of molecular adsorption usually destroys the surface states on Cu(110) [3,4] while the occupancy of the surface state does not seem to be a prerequisite for the surface to undergo reconstruction [24].

4.1. *RAS of the Au(110)-(1 × 2) surface*

The Au(110)-(1 × 2) surface has been the focus of a number of RAS studies in UHV [15,25,26,27,28,29,30] and at the solid/liquid interface [25,31,32,33]. At room temperature, the clean Au(110) surface is known to exhibit a (1 × 2) missing row reconstruction (Fig. 1c). The RA spectrum of the clean well-ordered Au(110)-(1×2) surface, shown in Fig. 3, is characterised by features at 1.80 eV, 2.52 eV, 3.52 eV and 4.50 eV (labelled A to D in Fig. 3).

Similar to the RA response of Cu(110), transitions involving surface-modified bulk bands contribute to the RA spectrum of Au(110). The RA peaks labelled C and D have been assigned to the transitions $E_F \to L_1^u$ and $L_2' \to L_1^u$, respectively [15]. The assignments are based upon a comparison between temperature-induced shifts in the energy of these RAS peaks and thermovariation optical spectroscopy results [22] of the temperature dependence of transition energies between the bands at L. There are indications of a contribution to the RA response in region A from transitions between surface states [25,26,28,30]. One such indication [30] is a decrease in RA intensity in region A upon exposure to air (Fig. 3). The contributions from surface-modified bulk bands to the RA response of both Cu(110) and Au(110) can be simulated using a 'three-phase' model that separates the system into vacuum, surface layer and substrate where the dielectric function of the surface layer is approximately proportional to the energy derivative of that of the substrate. The resulting simulated RA spectra agree well with the experimental results [14,15].

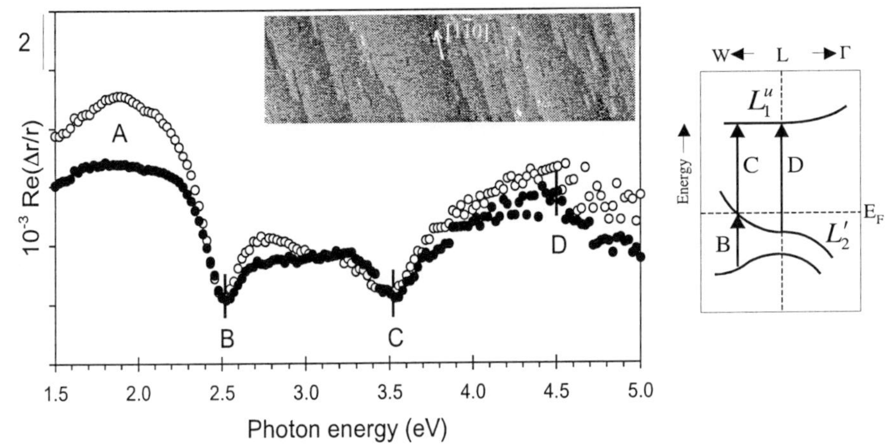

Figure 3. Left panel: RA spectra of Au(110)-(1 × 2) – the clean surface (open circles), and following exposure to ambient air (filled circles). Inset: STM image of the surface following exposure to ambient air shows (1 × 2) reconstructed terraces and monoatomic steps (100 × 30 nm). Right panel: Bands at L with corresponding transitions labelled.

5. RAS of molecular films at metal surfaces

As we have seen, the RA response of metals involves transitions between energy levels on a local atomic or molecular scale, however, RAS normally measures $(\Delta r/r)$ over a macroscopic area of typically $\sim$0.5 cm^2. The growth of molecular films on the macroscopically ordered (110) surfaces of metals offers a route to inducing macroscopic order into a molecular film due to the intrinsic structural anisotropy of these surfaces (Fig. 1). Such surfaces reduce the occurrence of differently oriented domains which can lead to a reduction or cancellation of the RAS signal by symmetry. RAS can also be used to monitor molecular assembly on disordered surfaces provided the molecules self assemble into a macroscopic ordered structure.

The Cu(110) surface has been the focus for RAS studies of molecular assembly in UHV. A wide range of molecules can bind to Cu(110) particularly if they contain a carboxylic acid (COOH) group that is available for interaction with the Cu surface. It is well known that the adsorption of mono-carboxylic acids onto Cu(110) leads to the deprotonation of the acid group into the carboxylate (OCO) functionality [34], which interacts with

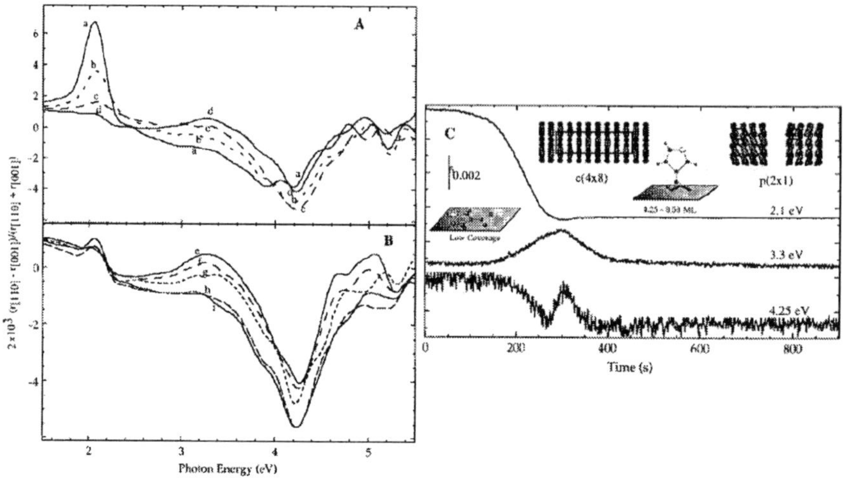

Figure 4. Left panel: RAS of increasing exposure (a–i) of 3TC on Cu(110). Right panel: single energy RA data. Changes observed in both panels relate to changes in molecular orientation with increasing coverage. Reprinted with permission from [3], copyright 1998 by the American Physical Society.

the Cu surface. At surface saturation, the carboxylate plane is oriented perpendicular to the Cu surface with the oxygen atoms interacting with adjacent close–packed Cu atoms along the $[1\bar{1}0]$ direction - a coordination known as bridging (see inset of Fig. 5). RAS has been found to be sensitive to the Cu-carboxylate interaction and has been used to investigate the formation of carboxylate–type molecular films on Cu(110). This was first demonstrated by a study of the adsorption of 3–thiophene carboxylic acid (3TC) on Cu(110) [3]. In this study, the combination of RAS, LEED, high-resolution energy loss spectroscopy and density functional calculations revealed the sensitivity of the RA response to the parallel and perpendicular orientation of the molecule at the surface.

Fig. 4 shows RAS of increasing exposure of 3TC on Cu(110). The changes observed in the data relate to changes in the orientation of the 3TC molecule from an initial flat lying geometry to a vertical geometry and then to a final tilted geometry as coverage is increased [3]. A similar study of the adsorption of 9–anthracene carboxylic acid onto the $p(2\times 1)$–O/Cu(110) surface [4] combined RAS, STM and infrared techniques to reveal the azimuthal orientation of the molecule on the surface and demonstrated the potential of RAS to determine molecular orientation by the observation of intra–molecular optical transitions.

The assembly of an ordered molecular layer at a surface raises the possibility of using such a layer to functionalise the surface. The carboxylate derivatives described above orient perpendicular to the surface at saturation and so the end groups of the molecule at the vacuum interface create a new surface. The use of dicarboxylic acids – molecules with acid groups at both ends – offers the chance to create an acid functionalised surface. This potential was realised using RAS and reflection absorption infra-red spectroscopy (RAIRS) to investigate the adsorption of (i) terephthalic acid [5] and (ii) oxalic acid [6] onto a stepped Cu(110) surface. The results of this work showed that the acid molecules adsorb as monocarboxylate species, bonding via the carboxylate group, to form a saturated layer of upright molecules that are terminated by acid groups - a functionalised surface. Fig. 5 shows the evolution of RA spectra with increasing terephthalate coverage up to the monolayer, and a diagram of the upright orientation found in the monolayer as determined from RAIRS data.

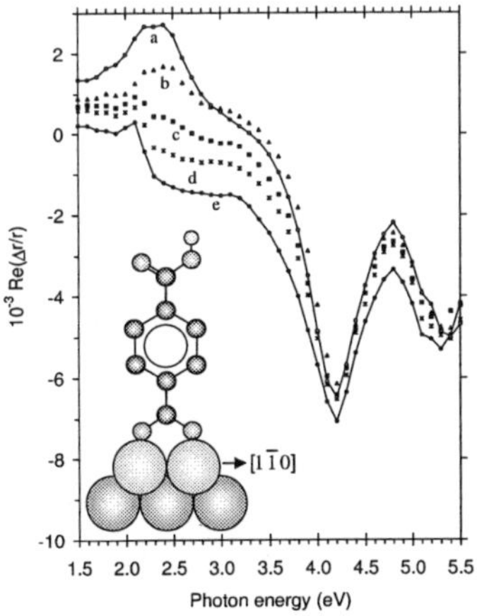

Figure 5. RAS data of the deposition of terephthalic acid onto Cu(110) at 340 K: (a) clean stepped Cu(110) surface, (b-e) build-up of the first terephthalate layer with (e) corresponding to the first layer. The inset shows the orientation of terephthalate in the first layer and the Cu–carboxylate bridging coordination. Reprinted from [7], copyright 2004, with permission from Elsevier.

The ability of RAS to investigate molecular assembly at the metal/liquid interface is of considerable technological importance. The Au(110) surface has been the focus of such studies due to the chemical inertness and stability of the Au surface relative to other metals, as demonstrated by the STM results of Fig. 3. Using an electrochemical cell, the adsorption of molecules can often be controlled via the applied potential. One example of such a system is the adsorption of pyridine species onto Au(110) [8,9]. Figure 6 shows the changes observed in the RA response upon adsorption of (a) pyridine, (b) 2,2'-bipyridine and (c) 4,4'-bipyridine. The pyridine species interact with Au through the N atoms and their expected orientations at the Au surface are shown in the figure. The changes in the RA profile upon adsorption suggest the molecules adopt an ordered structure with the differences between RA profiles for each pyridine species reflecting differences in molecular orientation at the Au surface.

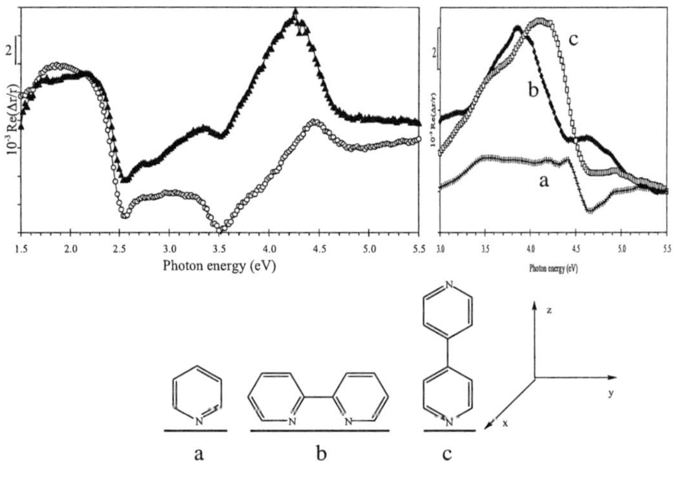

Figure 6. Upper left panel: RAS (RAS_{Au}) of Au(110) in 0.1 M $NaClO_4$ at 0.4 V vs. SCE (circles) and upon addition ($RAS_{molecule on Au}$) of 5 ml 4,4'-Bipyridine (triangles). Upper right panel: RA difference spectra ($RAS_{molecule on Au} - RAS_{Au}$) for the molecules a,b,c shown in the lower panel in their expected molecular orientation at the Au surface. Reproduced with permission of the authors.

RAS has been used to study ordering in molecular films that to assemble do not require an ordered substrate surface. The deposition of sapphyrin organic layers onto an Au-coated glass substrate by the Langmuir-Blodgett

(LB) technique has been examined using RAS [35]. RA spectra taken in air of the LB films were found to be characteristic of different layer thickness. STM studies of the same system [36] show preferential orientation of molecules into rows aligned along the direction of immersion of the sample into liquid during the LB deposition. Furthermore, the STM results show that the first layer of sapphyrin molecules orient nearly perpendicular to the substrate and subsequent layers orient nearly parallel to the substrate surface. The parallel orientation of molecules in the thick films produces a strong RAS signal near 500 nm ($\sim$2.5 eV) that is related to the in-plane optical anisotropy of the molecule. In a related study, RAS has been used to examine optical anisotropy of porphyrin LB films on quartz substrates [37]. These studies show the potential of RAS as a technique to investigate anisotropy in the ordering of complex molecules at surfaces.

6. Conclusions

The RAS technique has been used to monitor molecular assembly at metal surfaces in UHV and at the solid/liquid interface. For the Cu-carboxylate system, RAS has been shown to be sensitive to adsorbate–surface bonding and molecular orientation. The creation of functionalised surfaces presents the opportunity for molecular immobilisation or the growth of films with tailored physical and chemical properties. RAS in combination with STM and atomic force microscopy has a promising future as a probe of ordering and anisotropy in molecular films at the metal/vacuum and metal/liquid interfaces.

References

1. D.E. Aspnes, J.P. Harbison, A.A. Studna and L.T. Florez *J. Vac. Sci. Technol. A* **6** 1327 (1988)
2. D.S. Martin and P. Weightman *Surf. Interface Anal.* **31** 915 (2001)
3. B.G. Frederick *et al Phys. Rev. B* **58** 10883 (1998)
4. B.G. Frederick, J.R. Power, R.J. Cole, C.C. Perry, Q. Chen, S. Haq, Th. Bertrams, N.V. Richardson and P. Weightman *Phys. Rev. Lett.* **80** 4490 (1998)
5. D.S. Martin, R.J. Cole and S. Haq *Phys. Rev. B* **66** 155427 (2002)
6. D.S. Martin, R.J. Cole and S. Haq *Surf. Sci.* **539** 171 (2003)
7. D.S. Martin and P. Weightman *Thin Solid Films* **455-456** 752 (2004)
8. C.I. Smith, A.J. Maunder, C.A. Lucas, R.J. Nichols and P. Weightman *J. Electrochem. Soc.* **150** E233 (2003)
9. C.I. Smith, G.J. Dolan, T. Farrell, A.J. Maunder, D.G. Fernig, C. Edwards and P. Weightman *J. Phys.: Condens. Matter* in press

10. Ph. Hofmann, K.C. Rose, V. Fernandez, A.M. Bradshaw and W. Richter, *Phys. Rev. Lett.* **75** 2039 (1995)
11. K. Stahrenberg, Th. Herrmann, N. Esser, J. Sahm, W. Richter, S.V. Hoffmann and Ph. Hofmann *Phys. Rev. B* **58** 10207 (1998)
12. J.-K. Hansen, J. Bremer and O. Hunderi *Surf. Sci.* **418** L58 (1998)
13. D.S. Martin, A. Maunder and P. Weightman *Phys. Rev. B* **63** 155403 (2001)
14. L.D. Sun, M. Hohage, P. Zeppenfeld, R.E. Balderas-Navarro and K. Hingerl *Surf. Sci. Lett.* **527** L184 (2003)
15. D.S. Martin, R.J. Cole, N.P. Blanchard, G.E. Isted, D. Roseburgh and P. Weightman *J. Phys.: Condens. Matter* **16** S4375 (2004)
16. J.-K. Hansen, J. Bremer, and O. Hunderi, Phys. Stat. Sol. (a) **170**, 271 (1998).
17. N. Memmel *Surf. Sci. Rep.* **32** 91 (1998) and references therein
18. P. Roos, Ph.D. Thesis, Universität Bayreuth and IPP Report 9/108 (1995)
19. P. Straube, F. Pforte, T. Michalke, K. Berge, A. Gerlach and A. Goldmann *Phys. Rev. B* **61** 14072 (2000)
20. R. Schneider, H. Dürr, Th. Fauster and V. Dose *Phys. Rev. B* **42** 1638 (1990)
21. W. Jacob, V. Dose, U. Kolac, Th. Fauster and A. Goldmann *Z. Phys. B* **63** 459 (1986)
22. P. Winsemius, F.F van Kampen, H.P. Lengkeek and C.G. van Went *J. Phys. F.: Metal Phys.* **6** 1583 (1976)
23. F. Baumberger, Th. Herrmann, A. Kara, S. Stolbov, N. Esser, T.S. Rahman, J. Osterwalder, W. Richter and T. Greber *Phys. Rev. Lett.* **90** 177402 (2003)
24. D.S. Martin and P. Weightman *J. Phys.: Condens. Matter* **14** 675 (2002)
25. B. Sheridan, D.S. Martin, J.R. Power, S.D. Barrett, C.I. Smith, C.A. Lucas, R.J. Nichols and P. Weightman *Phys. Rev. Lett.* **85** 4618 (2000)
26. K. Stahrenberg, Th. Herrmann, N. Esser, W. Richter, S.V. Hoffmann and Ph. Hofmann *Phys. Rev. B* **65** 35407 (2001)
27. J.-K. Hansen, J. Bremer, L. Seime and O. Hunderi *Physica A* **298** 46 (2001)
28. D.S. Martin, N.P. Blanchard and P. Weightman *Surf. Sci.* **532-535** 1 (2003)
29. D.S. Martin, N.P. Blanchard and P. Weightman *Phys. Rev. B* **69** 113409 (2004)
30. G.E. Isted, D.S. Martin, L. Marnell and P. Weightman *Surf. Sci.* **566-568** 35 (2004)
31. V. Mazine, Y. Borensztein, L. Cagnon and P. Allongue *Phys. Status Solidi (a)* **175** 311 (1999)
32. V. Mazine and Y. Borensztein *Phys. Rev. Lett.* **88** 147403 (2002)
33. P. Weightman, C.I. Smith, D.S. Martin, C.A. Lucas, R.J. Nichols and S.D. Barrett *Phys. Rev. Lett.* **92** 199707 (2004)
34. M. Pascal, C.L.A. Lamont, M. Kittel, J.T. Hoeft, R. Terborg, M. Polcik, J.H. Kang, R. Toomes and D.P. Woodruff *Surf. Sci.* **492** 285 (2001)
35. C. Di Natale, C. Goletti, R. Paolesse, F. Della Sala, M. Drago, P. Chiaradia, P. Lugli and A. D Amico *Appl. Phys. Lett.* **77** 3164 (2000)
36. C. Goletti *et al Appl. Phys. Lett.* **75** 1237 (1999)
37. C. Goletti, R. Paolesse, C. Di Natale, G. Bussetti, P. Chiaradia, A. Froiio, L. Valli and A. D Amico *Surf. Sci.* **501** 31 (2002)

Study of solid/liquid interfaces by optical techniques

Y. Borensztein

Institute of NanoSciences of Paris, UMR CNRS 7588, Universities Paris VI and Paris VII, 4, place Jussieu, F-75252 Paris cedex 05, France. Email: borens@ccr.jussieu.fr

An overview of surface-sensitive optical techniques which are able to investigate the interfaces between a crystal and a liquid is given. Different examples are illustrated for several crystalline surfaces, which include use of linear optical techniques like Surface Differential Reflectance Spectroscopy, Reflectance Anisotropy Spectroscopy and Transform-Fourier Infra-Red on the one hand, and of non-linear optical ones like Second-Harmonic Generation and Sum-Frequency Generation on the other hand.

Introduction

Interfaces between solid materials and aqueous or non-aqueous liquids play a fundamental role in environment, in technology, in industrial applications. This is the case of metal corrosion or protection in presence of water, of cleaning of solid surfaces, of lubrication, of fabrication of thin layers or of new materials. Solid surfaces in liquid can also be modified by electrochemical techniques, which ensure control of charge transfer, adsorption of molecules or ions, chemical reactions, growth of films, crystalline reconstructions... It is therefore fundamental to be able to in-situ investigate the interface between a solid and a liquid. The observation of solid surfaces after transfer to a vacuum chamber, which opens the way to use classical surface technique, is not fully satisfying, because of the impossibility to control the modifications of the surface during immersion in the liquid and because of possible perturbations at the surface due to the changing the environments. Besides traditional electrochemistry techniques, and in-situ scanning probe microscopes, such as AFM and STM which have increasingly developed during the last decades and provide an atomistic view of the surface[1], few surface science techniques can be used for investigating surface/liquid interfaces. Due to the ability of photons to propagate in transparent media like liquids, surface-sensitive optical methods, which moreover are non-destructive, can be used to *in-situ* probe the solid/liquid interfaces. The aim of this article is to give an overview of the principal optical methods which are now currently used in this research area.

When a light beam impinges a solid surface (whatever the environment where it is placed), light propagates over nms or tens of nms inside the bulk. The information which can be gained by measuring, for example, the intensity of the reflected light, is dominated by the bulk contribution and carries only a very small contribution coming from the surface, of the order of 10^{-3} of the total signal[2]. It is therefore necessary to use specific techniques which allow one to get rid of bulk contributions, coming in our case both from the solid and the liquid. The first category of optical techniques, which are differential or modulation methods, provide differences in the signal upon change of a given parameter (applied potential, polarization of light, adlayer), which influences only the surface response but not the bulk one. Fourier-Transform-Infra-Red (FTIR) spectroscopy probes the vibrational properties of the interface[3,4], electroreflectance (ER)[5], Reflectance Anisotropy Spectroscopy (RAS)[2,6,7] and Surface Differential Reflectance Spectroscopy (SDRS)[2,8], working in the ultra-violet/visible range (UV-vis), inform on the electronic properties of the interface. A second kind of methods are the non-linear techniques, which are inherently sensitive to the interface only: Second-Harmonic-Generation (SHG)[9,10,11], used in UV-vis, and Sum-(Difference-)-Frequency-Generation (SFG and DFG)[9,10,12], used in infrared IR.

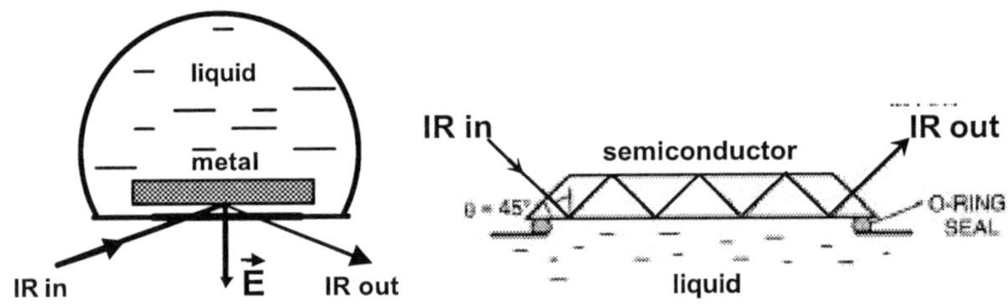

Fig. 1. Infrared set-up for a metal in oblique incidence (a) ; Multiple Infrared Reflection set-up for a semiconductor (b).

Infrared spectroscopy.

In-situ IR spectroscopy is now widely used to provide direct access to vibrations at the interface between a solid and a liquid, giving chemical fingerprints of the adsorbed species. The geometry differs as a function of the type of sample (Fig. 1). In the case of a metal, reflection with large angle of incidence is used. As the infrared light is slightly absorbed by aqueous liquids, it is necessary to work with a thin layer (1 μm) cell. Because reflection from a metal surface in IR is close to one, with change of phase for the electric field component parallel to the surface, only vibrations normal to the surface can be probed. On the contrary, for semiconductors which usually exhibit broad range of transparence in the IR, the Multiple Internal Reflection (MIR), or Attenuated Total Reflection (ATR), geometry is used. This permits one to enhance the sensitivity from the surface by using between 10 and 100 reflections, to work with regular cells which do not perturb the (electro-)chemical reactions at the surface, and to use different polarizations of light. An illustrative example of the possibilities of *in-situ* FTIR is the investigation of wet chemical etching of SiO_2 on silicon, which is an essential stage for the manufacture of silicon-based electronics. Fig. 2 shows the evolution of the Si-H vibration line for an oxidized silicon wafer, cut in the (111) direction and immersed in ammonium fluoride solution, measured in s- and in p- polarizations[13]. The absorption lines are broadened with respect to measurements in air performed on the same sample after it is removed from water, because of the interaction between solution molecules and surface hydrides. Nevertheless, comparison with the ex-situ measurements (not shown here) permits the authors to clearly assign the origin

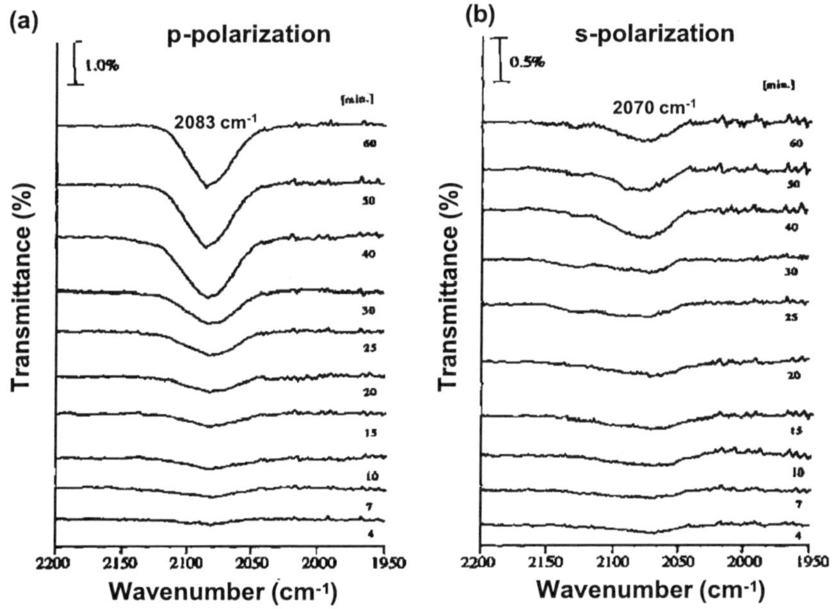

Fig.2. Time-dependent *in-situ* ATR-IR spectra of etching on Si(111) in a 25-times diltued ammonium fluoride solution (pH= 7.8). Reproduced from Ref. [13].

of the lines and to determine the etching process. The absorption peak at 2083 cm^{-1} is assigned to the Si-H stretching mode of ideal monohydride termination on the Si(111) terrace, characterized by a polarization perpendicular to the surface. The weak absorption after 4 min in both p- and s-pol. indicates the coexistence of ideal monohydride and other species, such as di- or tri-hydrides, at the first stage of etching. Both bands in p- and s-pol. keep growing during the removal of the oxide layer, reaching a maximum after 40 min, which indicates that the oxide is completely removed. After 40 min, the band in s-pol. decreases, which is explained by the removal of di- and tri-hydrides. The resultant Si(111) surface is slightly rough, with mainly monohydride species, either on terraces, visible only in p-pol. at 2083 cm^{-1}, and on steps, visible both in p- and in s-pol. at 2070 cm^{-1}. For comparison, the authors have also studied the etching process of Si(100) in the same conditions, not shown here: the spectra are now almost identical in s- and p- pol., which is shown to be due to an important roughness of the surface, which consists of (111) facets.

In order to have large intensity of the signal, most of the MIR experiments are performed with a large number of internal reflections, as it is the case in the previous example. This does not allow one to get direct information on the evolution of the SiO_2 film itself during etching, because of the complete absorption of the light by multiphonon bands in SiO_2. By using a shorter sample, which is more delicate because of a much smaller signal but reduces the number of reflections, Chabal and coworkers succeeded in monitoring the SiO_2 etching in HF, as illustrated in Fig. 3(a) where the disappearance of the TO and LO

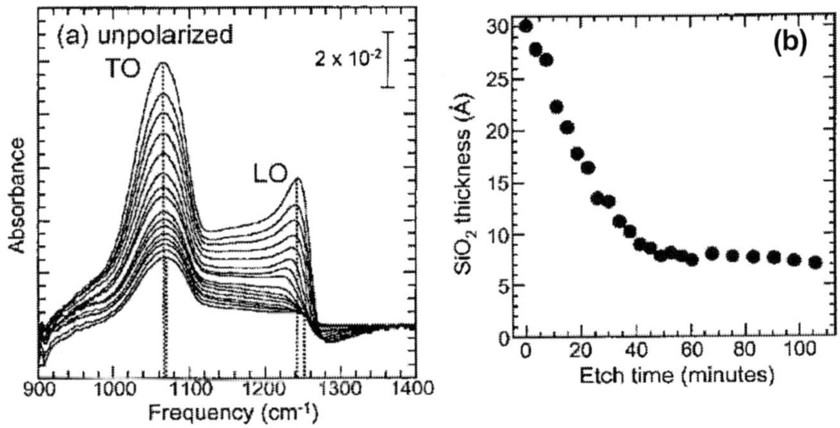

Fig. 3.(a): IR spectra acquired in-situ during etchnig of a 3 nm SiO2 layer on Si(100) in 0.05 % wt HF. (b): Plot of oxide thickness as a function of etching time. Reproduced from ref. [14]

phonons is observed as a function of time with unpolarized light.[14] The TO signal being proportional to the oxide thickness, the etching rate is then drawn in Fig. 3(b). The authors evidenced a clear transition, just after 40 min, from a linear etching regime to an apparent steady state of surface oxide thickness, which suggests a competition between surface oxidation and oxide removal.

Reflectance Anisotropy and Surface Differential Reflectance Spectroscopies.

Electroreflectance in the UV-visible range has been long-time used for investigating metal/electrolyte interfaces[5]. However, except in some cases where surface states could be clearly observed at the surface of well-ordered metals, and phase-transitions monitored by the evolution of these states[15], there is a strong influence in the ER spectra of bulk transitions, modified by the charging of the surface, and it is difficult to distinguish between purely surface effects and bulk effects. The Reflectance Anisotropy Spectroscopy, which has been widely used for the study of surface of semiconductor, has been applied during the past decade to investigate metal surfaces, in different environments[16,17,18]. RAS consists in measuring the anisotropy of normal-incidence reflectance of an anisotropic sample, by polarizing light along the two main axis of the surface. As UV-vis light is not absorbed by water, it can be used in a regular electrochemical cell. The use of electrochemistry permits to have an accurate control of the surface of the crystal[19].

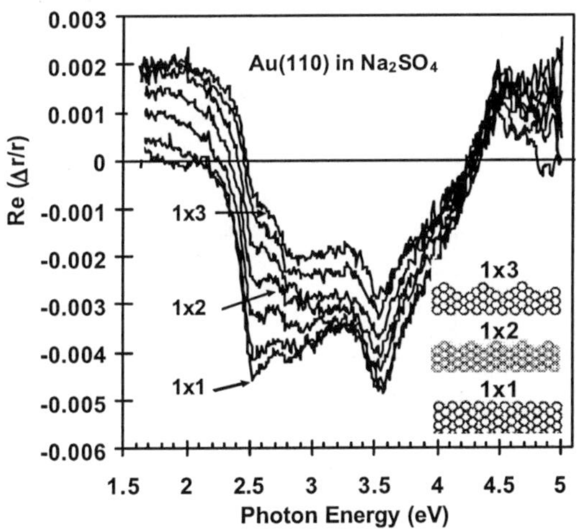

Fig. 4. RAS spectra for Au(110) in Na_2SO_4, defined by $\Delta r/r = (r_{[1\bar{1}0]} - r_{[001]})/r$. Applied potentials ranging from -0.6 V to 0.6 V (highest curve to lowest curve). The spectra drawn in thicker lines correspond to the 1x3, 1x2 and 1x1 reconstructions, obtained for potentials of -0.6, 0 and 0.6 V, respectively. Reproduced from Ref. [20].

RAS has been recently used to investigate the different reconstructions of Au(110) in electrolytes. As a function of the applied potential, the surface can be unreconstructed (1x1), or displays a one-missing row (1x2) reconstruction or a two-missing row reconstruction (1x3), as indicated in the inset of Fig. 4, which gives a profile view of the surfaces. The RA spectra, defined by $\Delta r/r = \left(r_{[1\bar{1}0]} - r_{[001]}\right)/r$ and plotted in Fig. 4, are different for the three surfaces and can be considered as the fingerprints of the reconstructions[20]. In-situ STM performed with the same samples in similar conditions confirms the reconstructions of the surface[21]. However, it must be pointed out that the origin of the anisotropy for Au(110) is not yet completely understood, and can be strongly dependent on the morphology of the surface, resulting in slightly different spectra, as shown by Weightman's group[22]. In spite of these differences, it can be seen on Fig. 4 that the intensity of the negative peak around 2.5 eV is very dependent on the applied potential. Consequently, the change of the intensity of this peak allowed us to monitor the transitions between the three surface phases, as shown in Fig. 5. Starting from a 1x3 reconstruction at negative potential, the surface undergoes a 1x3 $\rightarrow$ 1x2 transition followed by a 1x2 $\rightarrow$ 1x1 one, when the applied potential, therefore the surface charge density, reaches increasingly positive values. In the reverse half-cycle, no plateau can be seen, which is due to kinetic effects leading to a longer time for getting the reconstructions than for removing them. This observation is in good agreement with first-principle calculations which show that, the more negative the surface charge is, the more open the reconstruction of Au(110) is[15,23]. The continuous line in Fig. 5, which reproduces well the experimental points, is a result of a calculation based on the kinetic equations which describe the transitions between the different states[20].

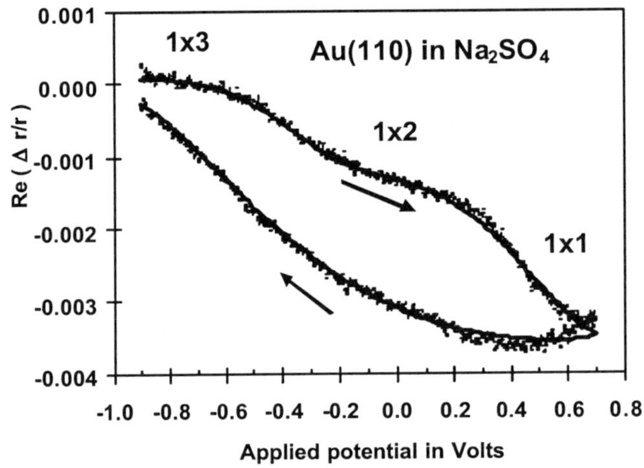

Fig.5. RA signal measured at 2.5 eV during potential cycle from -0.3 V to 0.6 V for Au(110) in H_2SO_4. Points: experiment ; continuous line: calculation. Reproduced from Ref. [20].

RAS can be also used to study the adsorption of molecules or ions in a liquid, or the growth of atom and molecule layer on a surface. A recent example is the study by RAS of the formation of pyridine layers grown on Au(110), which are shown to form ordered structures[24]. Another example is the electrochemical oxidation of Au(110). Although gold is considered as inert, oxidation can be forced in an electrolyte when applying a large positive potential. The combination of RAS with another surface-sensitive linear optical spectroscopy, the surface differential reflectance spectroscopy (SDRS), allowed us to monitor the oxidation of Au(110). SDRS provides the relative change of reflectance upon oxidation of the surface[28], defined by : $\Delta R/R = (R^{ox} - R^{clean})/R$, where R^{clean} and R^{ox} are the reflectances of the clean and oxidized Au surfaces respectively. Figure 6 gives the SDR spectra, measured in normal incidence with light polarized, either perpendicular or parallel to the atom rows, which are in the$[1\bar{1}0]$ direction. It is compared to the calculated curve expected for a thin oxide layer. Contrarily to the experimental spectrum along the [001] direction, which is close to the calculated one, there is a strong additional feature around 3.5 eV for the other polarization. This is caused by an optical absorption in the atom row direction, which is explained as due to polarizable bonds between oxygen and gold atoms, leading to a strongly anisotropic oxidized layer[25].

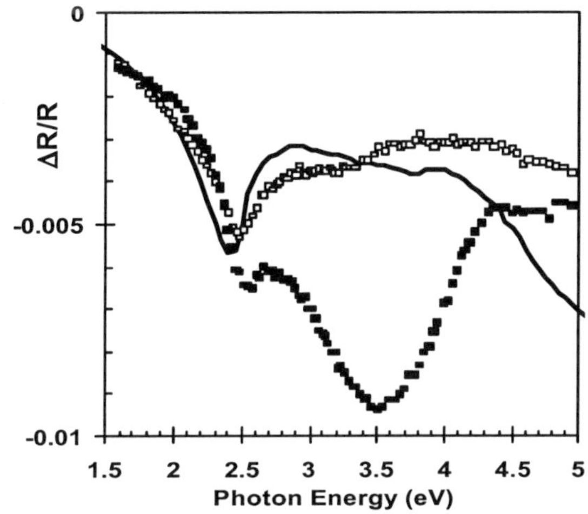

Fig. 6. Surface Differential Reflectance Spectra of Au(110) upon oxidation. Filled square: polarization parallel to the atom rows [1-10]. Empty squares: polarization perpendicular to the atoms rows. Continuous line: calculation for a thin layer of bolk oxide. Ref. [25]

Second-Harmonic Generation.

The basic mechanism of nonlinear optical spectroscopies, is that the interaction of light with matter is nonlinear if the intensity of incident light intensity is very high. In this case, light of double frequency (and of higher harmonics) is generated, which is called second harmonic generation (SHG). The second-harmonic polarization is written : $\vec{P}_i^{(2)}(\omega) = \chi_{ijk}^{(2)} \vec{E}_j(\omega)\vec{E}_k(\omega)$, where $\chi_{ijk}^{(2)}$ is the second-order non-linear susceptibility, which is a third-rank tensor. In the dipole approximation, SHG is forbidden in any centro-symmetric bulk material, such as fcc or hcp crystals. On the contrary, at the interface between the crystal and a liquid, where the symmetry is broken, second-harmonic can be generated. For this reason, SHG is inherently surface-sensitive, and probes only the very first atom layers[26]. The susceptibility tensor reflects the symmetry of the surface, which can be determined by analyzing the rotational anisotropy of the SHG intensity as a function of the azimuthal angle φ, obtained by rotating the sample around its surface normal[27].

For instance, in the case of a (111) surface, the SH electric field can be described as a function, which is periodic in φ with one isotropic term and three anisotropic terms, yielding the following expression for the SHG intensity: $I(2\omega) = |A + B\sin(\varphi) + C\sin(2\varphi) + D\sin(3\varphi)|^2$, where A is the isotropic contribution to the SH signal, and B, C, D are the one-fold, 2-fold and 3-fold terms[28]. An example is given in Fig. 7 where the SHG intensity of a Au(111) surface immersed in a H_2SO_4 electrolyte has been analyzed for two applied potentials[29]. At the more positive potential equal to 0.7 V, (a), the signal exhibits a perfect 3-fold symmetry, which corresponds to the surface symmetry of an unreconstructed (111) surface of a fcc crystal, with $I^{ps}(2\omega) \propto |\sin(3\varphi)|^2$. For a smaller potential of 0.2 V, (b), the angular rotation pattern is composed of a 3-fold symmetry and a 1-fold one: $I^{ss}(2\omega) = |B\sin(\varphi) + D\sin(3\varphi)|^2$, which is the indication of the formation of the so-called $22x\sqrt{3}$ (sometimes also called 1x23) superstructure, consisting of an hexagonal close-packed structure with an uniaxial compression of about 4% in the [110] direction[30]. The SHG signal, which reflects the symmetry of the surface, can therefore be used for monitoring reconstruction changes or formation of an adlayer as a function of the applied potential. The oxidation of Ag(111) has hence been monitored by SHG. Figure 8 shows the evolution of the two components which describe the SHG signal of Ag(111) : a 3-fold term coming from the very surface, and a zero-fold (isotropic) term which is less sensitive to the surface symmetry[31]. The increase of the

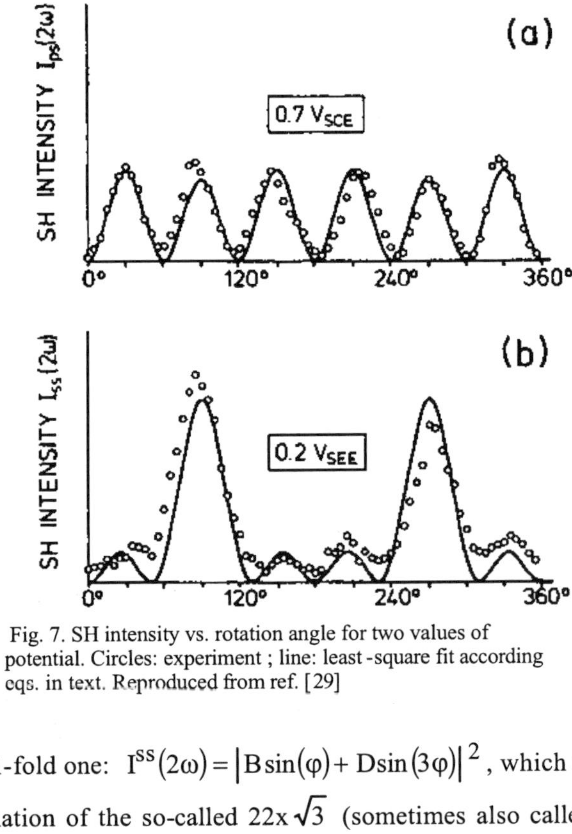

Fig. 7. SH intensity vs. rotation angle for two values of potential. Circles: experiment ; line: least-square fit according eqs. in text. Reproduced from ref. [29]

isotropic term as a function of the potential is shown to be due to a simple charging effect at the surface, which increases the SHG signal. On the contrary, the fast decrease at -0.1 V corresponds to the onset of bulk electrode oxidation. More interesting is the decrease of the 3-fold term at a smaller potential (-0.2 V), which is explained by a sub-monolayer surface oxide formation, which modifies the symmetry of the surface. SHG is hence shown to permit us to get a detailed understanding of surface and interface processes related to chemical reactions of a crystal immersed in a liquid.

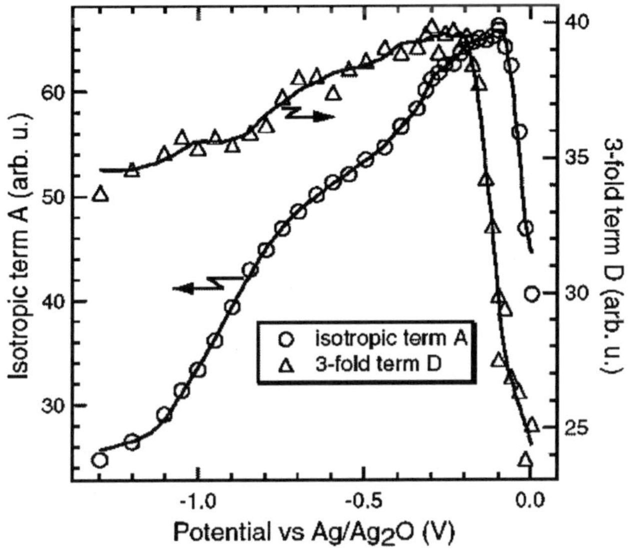

Fig. 8. SHG anisotropy coefficients of Ag(111) in 0.1 M NaF + 1mM NaOH. Plotted are the isotropic term A (circles) and the 3-fold term D (triangles) vs. potential. Reproduced from ref. [31]

Sum-Frequency Generation.

As in the case of SHG, the sum-frequency generation (SFG) is intrinsically sensitive to the interface, where the centro-symmetry of bulk is broken. Using two laser beams, one in the visible or UV range and the other one in the infra-red range, it permits one to probe the vibrations at the interface. The main interest with respect to FTIR spectroscopy is that no signal is obtained from molecules in the liquid, and that the signal is generated by the only adsorbed atoms or molecules[9,10]. However, as for FTIR, special care is needed because of the strong absorption of IR light by the liquid, and special cells have to be used. Fig. 9 shows a beautiful example of adsorption of hydrogen on Pt, during hydrogen evolution reaction (HER), whose understanding is fundamental for development of technologies for energy conversion and storage, like H_2/O_2 fuel cells. The SFG spectra were taken both in the underpotential (positive of the HER potential) and overpotential (negative) regions[9,32]. Two SFG resonances, at 1945 and 2020 cm^{-1} are visible in the underpotential region, which are due to Pt-H stretch of adsorbed atomic hydrogen on single platinum atoms. The splitting of this resonance arises from the interaction of the adsorbed hydrogen with water molecules of the electrolyte. At more negative potential, in the HER regime, an additional peak is observed at 1770 cm^{-1}, which appears only during HER. This peak is assigned to an active intermediate in the HER, which is the same on all different crystalline surfaces of Pt. It would consist of two hydrogen atoms bound on the same Pt atom, being a precursor for further evolution of H_2. This example was the first direct observation of H species at an electrode *during* HER, the interface being perturbed precisely by the evolution of gaseous hydrogen.

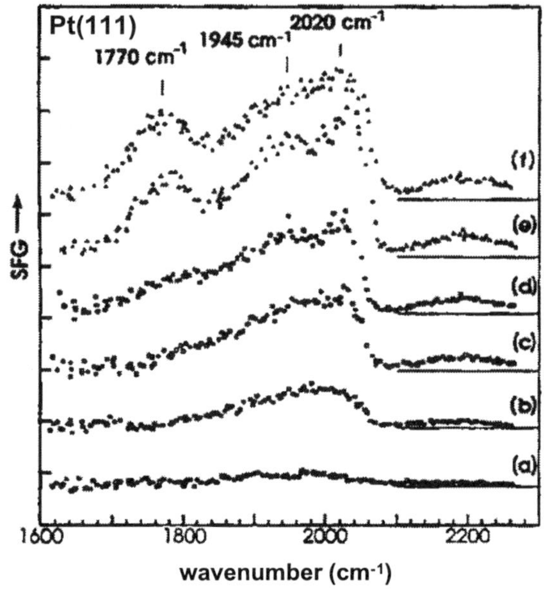

Fig. 9. SFG spectra for H on Pt(111) for decreasing potentials from 0.5 V (a) to -0.1 V (f). Reproduced from ref. [32].

Another remarkable property of SFG, is that is can be sensitive to the orientation of adsorbed molecules. When adsorbed on a metal surface, only vibrations normal to the surface can be probed by the IR light, like in FTIR. But, in addition, the orientation of the corresponding dipole of the molecule can be determined in some cases, when interference effects occur between the non-resonant SFG signal coming from the substrate surface and the IR-resonant signal from the molecule[33]. This is shown in Fig. 10, where SFG spectra of 4-cyanopyridine (4-CP) on Au(111) have been drawn for three different applied potentials, in the frequency range of the nitrile (CN) stretching mode. The comparison with *ab initio* calculations permits the authors to conclude that 4-CP is adsorbed perpendicular to the electrode surface, either through the pyridine nitrogen at negative potential or through the nitrile nitrogen at positive potential, and is adsorbed parallel to the surface, through the π electrons of the aromatic ring, in the intermediate potential, as shown in the scheme of Fig. 10.

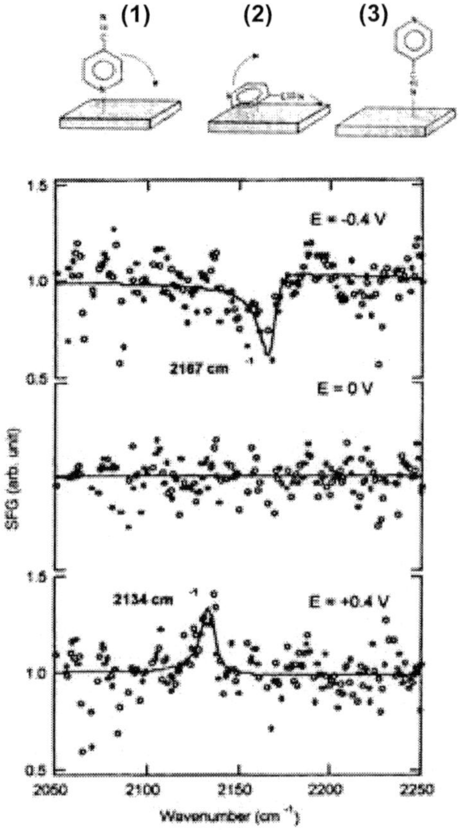

Fig.10. SFG spectra of 4-cyanopyridine molecules on Au(111) at three increasing potentials, in the CN stretching mode spectral range. Reproduced from Ref [33]

In conclusion, this review showed that various surface-sensitive optical techniques can be used efficently in order to probe the solid/liquid interface, and that they provide detailed information on the processes in play, such as surface reconstructions, adsorption of ions, atoms or molecules, passivation, formation of thin layers, etc.

---

[1] D.M. Kolb, Surf. Sci. 500, 722 (2002)
[2] Y. Borensztein, Surf. Rev. Lett. 7, 399 (2000)
[3] J.N. Chazalviel et al, J. Electroan. Chem. 509, 108 (2001)
[4] Y. Chabal and K. Raghavarachi, Surf. Sci. 502-503, 41 (2002)
[5] D.M. Kolb, in *Spectroelectrochemistry, Theory and Practice*, p.87-188, R.J. Gale ed., Plenum Press, New-York (1988)

[6] see Special Issue on *Linear Optics at Surfaces and Interfaces,* J. Phys. Cond. Mat. 16, pp. S4243-S4402 (2004)
[7] R. Del Sole, in *Photonic probes of Surfaces*, Ed. P. Halevi, Elsevier Science, chap. 4 (1995)
[8] P. Chiaradia and G. Chiarotti, in *Photonic probes of Surfaces*, Ed. P. Halevi, Elsevier Science, chap. 3, (1995)
[9] A. Tadjeddine and A. Peremans, in *Spectroscopy for Surface Science*, R.J.H. Clark and R.E. Hester ed., John Wiley & Sons, p.159-217 (1998)
[10] C.T. Williams and D.A. Beattie, Surf. Sci. 500, 545 (2002)
[11] E.D. Mishina et al, Appl. Phys. B74, 765 (2002)
[12] A. Tadjeddine et al, Phys. Stat. Sol. (a) 175, 89 (1999)
[13] M. Nakamura, M.B. Song and M. Ito, Electrochimica Acta 41, 681 (1996)
[14] K.T. Queeney et al, J. Phys. Chem. B 105, 3903 (2001)
[15] D.M. Kolb, Progress in Surface Science 51, 109 (1996)
[16] Y. Borensztein et al, Phys. Rev Lett. 71, 2334 (1993)
[17] W.L. Mochan et al, Physica A207, 334 (1994)
[18] Ph. Hofmann et al, Phys. Rev. Lett. 75, 2039 (1995)
[19] T.E. Furtak, Surf. Sci. 299/300, 945 (1994)
[20] V. Mazine and Y. Borensztein, Phys. Rev. Lett. 88, 147403 (2002)
[21] V. Mazine et al, Phys. Stat. Sol. (a) 175, 311 (1999)
[22] B. Sheridan et al, Phys. Rev. Lett. 85, 4618 (2000) ; P. Weightman et al, Phys. Rev. Lett. 92, 199707 (2004).
[23] K.M. Ho and K.P. Bohnen, Phys. Rev. Lett. 59, 1833 (1987)
[24] C.I. Smith et al, J. Elect. Soc. 150, E233 (2003)
[25] Y. Borensztein and V. Mazine, to be published
[26] G.A. Reider and T.F. Heinz, *Photonic probes of Surfaces*, Ed. P. Halevi, Elsevier Science, chap. 9 (1995)
[27] J.E. Sipe, D.J. Moss and H.M. van Driel, Phys. Rev. B35, 1129 (1987)
[28] B. Pettinger et al, in *Interfacial Electrochemistry*, ed. A. Wieckowski (Marcel Dekker, NY 1999), pp. 373-404 ; M. Danckwerts et al, Appl. Phys. B74, 635 (2002)
[29] G. Lüpke et al, Phys. Rev. B41, 6913 (1990)
[30] J.V. Barth et al, Phys. Rev B42, 9307 (1990)
[31] M. Danckwerts et al, Appl. Phys. B74, 635 (2002)
[32] A. Peremans and A. Tadjeddine, J. Chem. Phys. 103, 7197 (1995)
[33] O. Pluchery et al, Phys. Chem. Chem. Phys. 3, 3343 (2001)

SURFACE PREPARATION OF CU(110) FOR AMBIENT ENVIRONMENTS

G.E. ISTED, N.P. BLANCHARD AND D.S. MARTIN*

Department of Physics and Surface Science Research Centre,
University of Liverpool,
Liverpool, L69 3BX, England
**E-mail: davidm@liverpool.ac.uk*

The preparation of metal surfaces that are smooth and flat over micron length scales is desirable for high resolution scanning probe microscopy (SPM) studies of adsorbed molecules. Standard polishing finishes are often found to exhibit considerable roughness and damage including scratches when investigated by SPM. We have prepared by means of UHV technology a Cu surface that when in ambient air is considerably smoother than conventional polished surfaces. The RA response in the region 3.0 – 5.0 eV of this Cu surface can be interpreted in terms of surface–modified bulk bands. Future experiments in our research group will utilise such surfaces for studies of biomolecular adsorption.

1. Introduction

The preparation of metal surfaces that are smooth and flat over micron length scales is desirable for high resolution scanning probe microscopy (SPM) studies of adsorbed molecules such as proteins and DNA. Standard polishing methods commonly applied to metal surfaces appear to yield a mirror–smooth finish, however, such finishes are often found to exhibit considerable roughness and damage including scratches when investigated by SPM. An atomic force microscopy (AFM) image of a polished Cu(110) surface is shown in Fig. 1(i). Clearly such a surface morphology is not suitable for studies of molecular adsorption. Standard surface cleaning cycles of Ar ion bombardment and annealing in an ultra-high vacuum (UHV) environment, followed by annealing at high temperature, is known to generate surfaces of atomically flat terraces and monoatomic height steps [1,2]. If a smooth morphology could be preserved upon removing surfaces from UHV into ambient conditions then such a surface would satisfy one requirement for high resolution SPM imaging. In this article we give an overview of a method for the preparation of a copper surface that exhibits a relatively smooth morphology in ambient conditions. We characterise this surface using SPM, reflection anisotropy spectroscopy (RAS) and x-ray photoelectron spectroscopy (XPS).

2. Experimental details

A Cu(110) crystal was polished to a final grade of 0.25 micron and following AFM analysis the specimen was transferred into a UHV chamber for cleaning. Standard cleaning cycles of Ar ion bombardment and annealing produced a clean and atomically ordered surface as determined by x-ray photoelectron spectroscopy (XPS) and low energy electron diffraction (LEED), respectively. The specimen was then annealed to $\sim$ 1060 K - near the roughening temperature of Cu. Following cooling to room temperature, a well–ordered, stepped surface was observed. Steps were oriented in the [001] direction and terraces were typically $\sim$ 5 nm in width. The stepped surface was exposed to ambient air via a sample transfer mechanism allowing isolation of the specimen from the main chamber while the pressure is raised to 1 atm using high purity nitrogen. The specimen is then exposed to ambient air for $\sim$ 5 min before being transferred back into UHV for subsequent analysis.

RAS[3,4] is defined as:

$$\frac{\Delta r}{r} = \frac{2(r_{[1\bar{1}0]} - r_{[001]})}{r_{[1\bar{1}0]} + r_{[001]}} \tag{1}$$

Spectra of the real part of the complex RA were recorded between 1.5 – 5.0 eV.

3. Surface morphology

Fig. 1 shows the morphology of (i) the polished surface in ambient air, (ii) detail of the clean and stepped surface in UHV and (iii) detail of the surface (ii) following exposure to ambient air as imaged in UHV by STM. Fig. 1 (iv) shows a larger scale AFM image of (iii) obtained with the specimen in ambient air. Comparing (iv) with (i) it is clear that the morphology of the surface has improved dramatically, particularly in terms of decreased roughness. Note also the lack of any surface damage such as scratches in (iv). Having prepared a relatively smooth surface morphology we now characterise its chemical and optical properties.

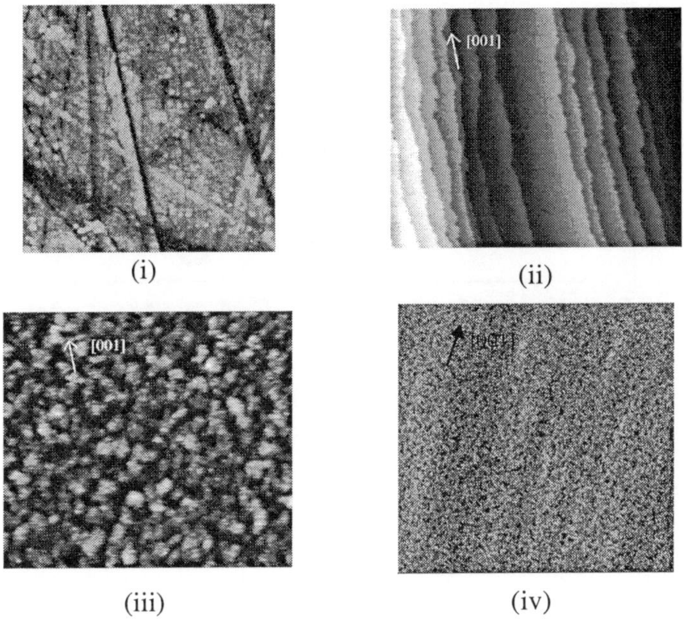

Figure 1. (i) AFM of the polished Cu surface in ambient air $(2 \times 2)\mu m$. STM of (ii) the clean and stepped surface in UHV (200×170)nm and (iii) the surface (ii) following exposure to ambient air (150×135)nm. (iv) AFM image of (iii) obtained with the specimen in ambient air $(3 \times 3)\mu m$.

4. Chemical properties

XPS data of the Cu $2p_{1/2}$, Cu $2p_{3/2}$ and Cu L_3VV peaks of the air exposed surface was similar to previous results [5]. The XPS data indicate the presence of a cuprous oxide (Cu_2O) overlayer.

5. Optical properties

RA spectra of the clean, stepped surface and the effect on this surface of exposure to ambient air are shown in Fig. 2. Differences between the two spectra are due to both the differences in morphology (Fig. 1) and chemistry (clean Cu or cuprous oxide) of the two surfaces. The RA spectrum of the clean, stepped surface has been discussed in more detail by Martin *et al* [1]. The spectrum of the air–exposed surface is broadly similar to that observed by Stahrenberg *et al* for an air–exposed surface created from a non-stepped Cu(110) surface [6].

The RAS of clean Cu(110) in the region 4–5 eV has been attributed

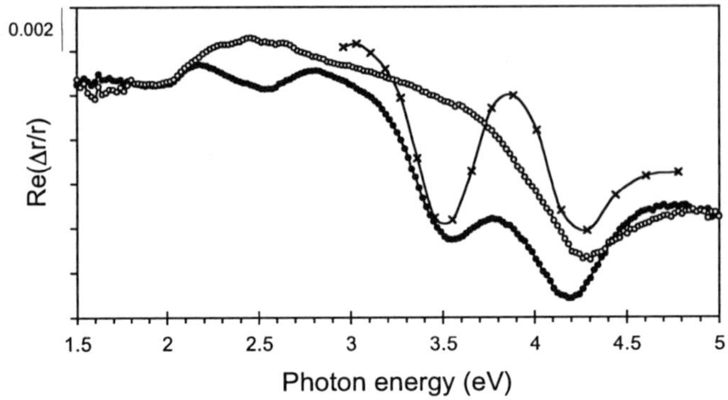

Figure 2. RA spectra of the clean, stepped surface in UHV (open circles) and this surface following exposure to ambient air (filled circles). Simulated RA spectrum between 3.0 – 5.0 eV (crosses) of the air exposed surface as outlined in the text.

[7] to transitions involving surface–modified bulk bands near E_F at the L symmetry point. It follows from this interpretation that this region of the spectrum may be simulated based upon knowledge of the bulk dielectric function. We simulate the RA response of the air-exposed surface using a three-phase model [8] of vacuum, surface layer and substrate with each phase having a distinct dielectric function. In the thin film limit we have:

$$\frac{\Delta r}{r} = -\frac{4\pi i d}{\lambda}\frac{\Delta\epsilon_s}{\epsilon_b - 1} \tag{2}$$

where ϵ_b is the substrate dielectric function and $\Delta\epsilon_s$ is the dielectric anisotropy in the surface layer. In our simulation ϵ_b'' (we use the notation $\epsilon = \epsilon' + i\epsilon''$) was determined using spectroscopic ellipsometry data of a Cu(110) crystal in air [9] and hence takes into account the presence of the cuprous oxide overlayer. In the surface layer the dielectric function is approximately proportional to the energy derivative of ϵ_b [10]. For Cu(110) there are different energy and broadening shifts for $[1\bar{1}0]$ and $[001]$ polarised light leading to:

$$\frac{\Delta r}{r} = -\frac{4\pi i d}{\lambda}\frac{(\Delta E_g + i\Delta\Gamma)}{\epsilon_b - 1}\frac{d\epsilon_b}{dE} \tag{3}$$

where ΔE_g and $\Delta\Gamma$ are the relative shifts in gap energies and linewidths for the two polarisations, and E is the photon energy. The simulated RA spectrum in the region 3.0–5.0 eV obtained using the parameters $\Delta E =$ -0.1 eV and $\Delta\Gamma =$ -0.1 eV is plotted in Fig. 2. Good agreement is found between the simulation and experimental results in terms of both the general shape and the energies of the RAS peaks. This result indicates that the RA response of the air–exposed surface between 3.0–5.0 eV is dominated by surface–modified bulk effects.

6. Conclusions

We have prepared by means of UHV technology a Cu surface that when in ambient air is considerably smoother than conventional polished surfaces. Following exposure of clean Cu(110) to ambient air, a cuprous oxide (Cu_2O) overlayer is formed. The RA response in the region 3.0 – 5.0 eV of the air–exposed surface can be interpreted in terms of surface–modified bulk bands. Future experiments in our research group will utilise this surface for studies of biomolecular adsorption.

7. Acknowledgment

We acknowledge the financial support of the UK EPSRC. We thank P. Weightman for access to UHV facilities.

References

1. D.S. Martin, A. Maunder and P. Weightman *Phys. Rev. B* **63** 155403 (2001)
2. G.E. Isted, D.S. Martin, L. Marnell and P. Weightman *Surf. Sci.* **566-568** 35 (2004)
3. D.E. Aspnes, J.P. Harbison, A.A. Studna and L.T. Florez *J. Vac. Sci. Technol. A* **6** 1327 (1988)
4. D.S. Martin and P. Weightman *Surf. Interface Anal.* **31** 915 (2001)
5. S.K. Chawla, B.I. Rickett, N. Sankerreman, J.H. Payer *Corros. Sci.* **33** 1617 (1992)
6. K. Stahrenberg, Th. Herrmann, N. Esser and W. Richter *Phys. Rev. B* **61** 3043 (2000)
7. L.D. Sun, M. Hohage, P. Zeppenfeld, R.E. Balderas-Navarro and K. Hingerl *Surf. Sci. Lett.* **527** L184 (2003)
8. J.D.E. McIntyre and D.E. Aspnes *Surf. Sci.* **24** 417 (1971)
9. N.P. Blanchard, PhD Thesis, University of Liverpool (2004).
10. U. Rossow, L. Mantese and D.E. Aspnes *J. Vac. Sci. Technol. B* **14** 3070 (1996)

MICRO-RADIOGRAPHS STORED IN LITHIUM FLUORIDE FILMS SHOW STRONG OPTICAL CONTRAST WITH NO TOPOGRAPHICAL CONTRIBUTION

A. USTIONE, *A. CRICENTI

Istituto di Struttura della Materia – C.N.R.,
via fosso del cavaliere 100, 00133 Rome, Italy
**Dep. of Physics and Astronomy, Vanderbilt University,*
Nashville, TN 37212 USA.

F. BONFIGLI, F. FLORA, A. LAI, T. MAROLO, R. M. MONTEREALI, G. BALDACCHINI

ENEA, UTS, Advanced Physical Technologies,
C.R. Frascati, Via E. Fermi 45, 00044 Frascati, Rome, Italy

A. FAENOV, T. PIKUZ

MISDC of VNIIFTRI Mendeleevo,
Moscow region, 141570, Russia

L. REALE

Università de L'Aquila e INFN, Physics dept.,
Coppito, L'Aquila, Italy

Using films of Lithium Fluoride as image detectors, and a laser-plasma source to produce soft X-rays and Extreme Ultraviolet radiation, micro-radiographs of biological material have been produced, by the creation of permanent point defects in the Lithium Fluoride lattice which are known as Color Centers. These micro-radiographs can be read exploiting the Optically Stimulated Luminescence of the Color Centers. In these paper we present the results of measures conducted with a Scanning Near-Field Optical Microscope, that show optical details with no topographical contribution at all.

The SNOM (Scanning Near-field Optical Microscope) is a member of the large family of Scanning Probe Microscopes[1-5]. It works keeping a small aperture at a low distance from a surface that is illuminated by a light source, and collecting the non-propagating component of the radiation which decays exponentially going away from the sample's surface. This electromagnetic radiation is the near-field that contains information of the optical properties of the sample, with a spatial resolution lower than the source wavelength and so overcoming the diffraction limit.

The used SNOM is an aperture reflection system working in the "collection mode"[6]. It uses uncoated tip, that we produce tapering an optical fiber, to collect the near field generated illuminating the sample with laser light. The system uses a piezoelectric to make the tip dither, while a bimorph piezoelectric detects the shear-force between the vibrating tip and the surface, feeding a feedback loop that repositions the sample under the tip: in this way the aperture is kept at a small constant distance from the surface during the whole scan. The electronic to control the scanning unit, the data acquisition system and the image elaboration are fully described elsewhere[7-9]. The system simultaneously acquires topographical and optical images to recognize possible artifacts.

Lithium fluoride (LiF) is a well-known radiation-sensitive material, that can be grown as a thin film on different substrates[10, 11]. Recently, LiF was used as image detector for X-ray micro-radiography and X-ray microscopy[12]. In fact extreme ultraviolet (EUV) radiation and soft X-rays locally produce permanent point defects[13] in LiF known as Color Centers (CCs), and some of this CCs emit visible light under selective optical excitation. Biological sample, placed in contact with the LiF film, were exposed to EUV and soft X-rays (20eV - 1500eV) produced by a laser-plasma source[14-16] and then removed. The result is an image that is stored in the film and that can be read illuminating the film itself with a blue light ($\lambda = 458$ nm) and collecting the yellowish light emitted by F_3^+ and F_2 CCs which posses two broad emission bands peaked around 535 nm and 680 nm, respectively[11]. We used a long-pass filter with a cutoff wavelength of 510 nm to collect only the integrated fluorescence of the F_3^+ and F_2 CCs.

We present SNOM images of a micro-radiograph of a mosquito (*Diptera*) wing. The film was grown evaporating LiF on a silicon substrate heated at a temperature of 250 °C. The thickness of the film is 120 nm. Then the mosquito wing was put over it and the sample was irradiated using EUV radiation and Soft X-rays (20 eV < hv < 1500 eV) by a laser-plasma source. Finally the wing was removed and the stored image in LiF film was observed by the SNOM system. As the silicon isn't transparent in the visible region of the spectrum, this sample can be studied only with a SNOM system that is able to work in reflection mode.

SNOM images of the micro-radiograph of a mosquito wing are shown in Figs. 1a and 1b. The smooth surface of the LiF film, that has a typical roughness of ~ 20 nm[17], is covered by particles of various nature (debris, dust, etc.). On the contrary, the corresponding fluorescence image, shown in Fig. 1b, clearly reveals the details of a small portion of the wing: little hairs, having width between 0.5 μm and 1 μm and length between 6 μm and 10 μm, are spread all over the wing. Due to the hair higher density, they absorbed more soft X-rays than the rest of the wing and so they appear darker in the micro-radiograph. The comparison of the topographical and the optical images reveals how the latter are free from instrumental artifacts: in fact the observed contrasts are not related to topographical structures but only due to the different doses of soft X-rays that reached the LiF film having passed the biological sample. It's worth noting that the black areas in the optical image, don't correspond to an absolute lack of fluorescence signal. In fact a layer of ~ 10 μm of biological material is necessary to completely stop soft- X-rays[16], and the hairs are not so thick. On the contrary in Fig 2b the micro-radiograph includes a portion of the wing skeleton that is the thickest part of the wing. Here we can see a large black area that corresponds to the complete absence of fluorescence signal.

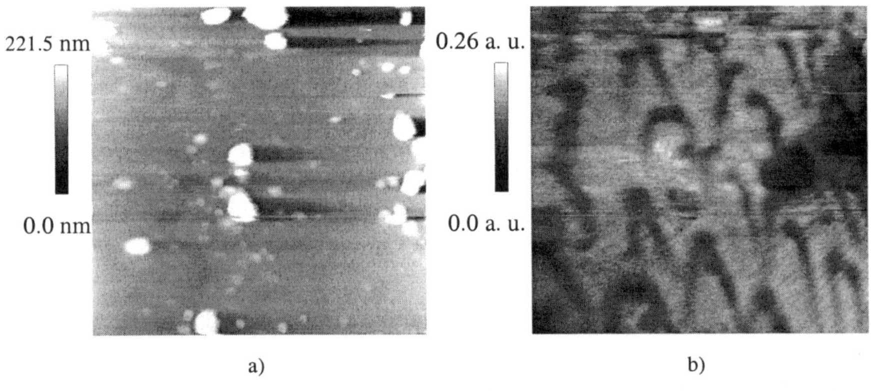

Fig. 1*:* a) 40 μm × 40 μm topographical image showing the LiF surfacc and particles of various nature laying on it. b) Corresponding SNOM fluorescence image. The micro-radiograph shows details of the wing and is free from topographical contribution or instrumental artifacts.

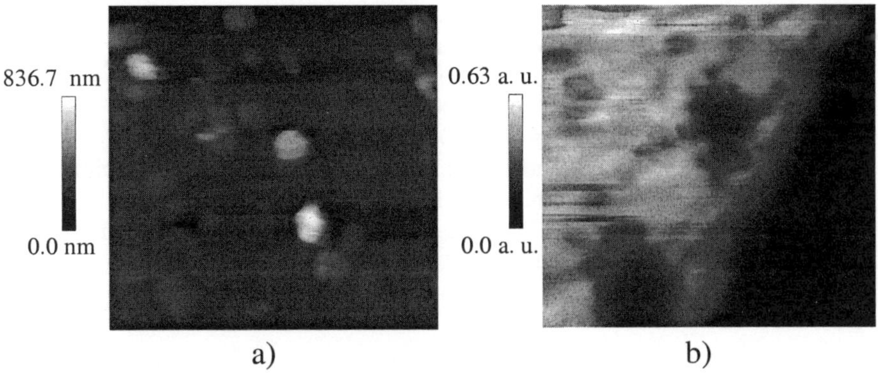

Fig. 2: a) 40 μm × 40 μm topographical image. b) SNOM fluorescence image of the same area. The presence of the thick wing skeleton stopped the soft X-rays and we observe the dark area corresponding to the absence of CCs in the LiF film.

We have presented a recent application in which a film of lithium fluoride was used as a detector for soft X-rays micro-radiography. The stored biological image is read by observing the optically stimulated visible luminescence of the active color centers, efficiently produced by the soft X-rays. We demonstrated that the used aperture reflection SNOM in "collection mode" can be successfully used to observe details of micro-radiographs stored in LiF detector, and that the obtained images are free from topographical or instrumental artifacts.

REFERENCES

1. D. W. Pohl, U. C. Fischer, U. T. Durig, J. Microsc. **152**, 853 (1998)
2. G. Binnig, C. F. Quate, C. Gerber, Phys. Rev. Lett. **56**, 930 (1986)
3. D. W. Pohl, D. Courjon (Eds.), Near Field Optics, NATO ASI series E-242, Kluwer Academic Publisher (1993)
4. E. Tuncel, *et al.*, Phys. Rev. Lett. **70**, 4146 (1993)
5. C. Coluzza, *et al.*, Phys. Rev. B **46**, 12834 (1992)
6. J. A. Cline, M. Isaacson, Ultramicroscopy **57**, 147 (1995)
7. C. Barchesi, *et al.*, Rev. Sci. Instrum. **68**, 3799 (1997)
8. A. Cricenti, R. Generosi, Rev. Sci. Instrum. **66**, 2843 (1995)
9. A. Cricenti, *et al.*, Rev. Sci. Instrum. **69**, 3240 (1998)
10. R. M. Montereali *Point Defects in Thin Insulating Films of Lithium Fluoride for Optical Microsystems,* in Handbook of Thin Film Materials, H. S. Nalwa ed., Vol.3: Ferroelectric and Dielectric Thin Films, 2002, Academic Press, Ch.7, pp. 399-431

11. R. M. Montereali, *et al.*, Thin Solid Films **196** 75-83 (1991)
12. G. Baldacchini, *et al.*, J. Nanosci. Nanotech. **3**, 483 (2003)
13. G. Baldacchini, *et al.*, Appl. Phys. Lett. **80**, 4810 (2002)
14. S. Bollanti, *et al.*, Il nuovo Cimento **20D**, 1685 (1998)
15. P. Albertano, *et al.*, J. Microsc. **187**, 96 (1997)
16. D. Attwood, Soft X-rays and Extreme Ultraviolet Radiation: Principles and Applications, Cambridge University Press, 1999
17. R. M. Montereali, *et al.*, Nuclear Instruments and Methods in Physics Research B **166-167** (2000) 764-770

Metal nanofilms studied with infrared spectroscopy

Gerhard Fahsold, Andreas Priebe, and Annemarie Pucci[a]

Kirchhoff Institute of Physics, Heidelberg University, Germany,

Andreas Otto

Institut für Physik der kondensierten Materie, Universität Düsseldorf, Germany

Metal films with thickness in the nanometer range are optically transparent. In the IR range their transmittance may show both the Drude-type behaviour of coalesced islands and the tail of the plasmon absorption of single islands. Therefore, IR transmittance spectroscopy is a sensitive tool for *in-situ* studies of metal-film growth on insulating substrates and of the film conductivity. With IR transmittance spectroscopy the in-plane film conductivity and its correlation to the film-growth process can be determined without electrical contacts. Adsorbate induced changes can be observed well. Their analysis may give insight into the adsorbate-metal bonding. Depending on the film's roughness the IR lines of adsorbate-vibration modes may be strongly modified because of their interaction with electronic excitations of the film. The atomic roughness of cold-condensed metal films produces additional IR activity: strong IR activity of Raman lines of centrosymmetric adsorbate molecules is observed in those cases where the adsorbate has states close to the Fermi level.

Introduction

Infrared spectroscopy (IRS) enables the study of free charge carriers and of atomic vibrations within one experiment. Metal films with thicknesses of few nanometers (nanofilms) are transparent and, therefore, their conductivity can be studied easily via transmittance measurements. For such studies no electrical contacts are necessary and the macroscopic surface defects do not disturb the conductivity studies. For nanofilms the adsorbate induced conductivity changes are much more pronounced than for thick films. They may significantly increase or decrease the nanofilm conductivity.

Several other interesting properties and effects are related to metal nanofilms. They are, for example, surface enhanced IR absorption (SEIRA)[1,2,3] and roughness induced catalytic activity[4], which can be studied in situ also with IRS[5]. Certain defect sites give rise to the appearance of additional IR lines belonging to possibly active species as it will be shown below for C_2H_4 adsorption. Concerning SEIRA which increases the intensity also of such defect lines, it is interesting that close to percolation of island-like films the interaction of vibrations with surface plasmons leads to SEIRA with asymmetric lines.[2]

[a] E-mail: pucci@kip.uni-heidelberg.de

Basics

Optical properties of metals in the IR are determined by collective oscillations of free charge carriers (plasmons). Therefore, in the far to middle IR the Drude-type dielectric function $\varepsilon(\omega)$ is valid for the most metals:

$$\varepsilon(\omega)-\varepsilon_{\infty}=-\frac{\omega_{\mathrm{p}}^{2}}{\left(\mathrm{i}\omega/\tau+\omega^{2}\right)}. \qquad (1)$$

Hereby it is important to note, that the parameter plasma frequency (ω_{P}) follows from the band structure of the metal and may be different from simple estimates. Going to higher circular frequencies ω (from the middle IR) both the parameters ω_{P} and the lifetime $\tau=\omega_{\tau}^{-1}$ become frequency dependent[6] because of many-particle effects and energy dependent scattering. In ultrathin films, the resistivity is dominated by surface scattering, which decreases absorption in the IR, and therefore, allows the application of local optics.[7]

Metal-film spectra

At normal incidence of light the transmittance $T_{\mathrm{film/substrate}}$ of an ultrathin metal film (thickness d << IR wavelength inside film) relative to the transmittance $T_{\mathrm{substrate}}$ of the pure transparent substrate (with refractive index $n_{\mathrm{substrate}}$) approximately is

$$\frac{T_{\mathrm{film/substrate}}}{T_{\mathrm{substrate}}}\approx 1-\frac{2\cdot d\cdot\omega\cdot\mathrm{Im}\varepsilon_{\mathrm{film}}(\omega)}{c\cdot(1+n_{\mathrm{substrate}})}=1-\frac{2\cdot d\cdot\mathrm{Re}\sigma_{\mathrm{film}}(\omega)}{c\cdot(1+n_{\mathrm{substrate}})\varepsilon_{0}}. \qquad (2)$$

Here ε_0 is the vacuum dielectric constant and c is the velocity of light. Eq. 2 should make obvious that the IR dynamic conductivity σ_{film} can be measured with IR transmittance spectroscopy. Then the dc conductivity is the limit of σ_{film} for $\omega\rightarrow 0$. It is important to note, that in equation (2) only the conductivity parallel to the film is involved and that for more precise data analysis the full Fresnel coefficients for thin-film optics have to be taken into account: For continuous films we fitted the IR spectra on the basis of the Drude model and under consideration of size effects that may modify charge carrier scattering and plasma frequency.[7,8,9] In this way it is possible to find out the thickness where the film has reached bulk conductivity. For example metal-film growth on UHV cleaved MgO(001) at 300 K by physical vapor deposition with rates of about few Å per minute : Iron films reach the bulk conductivity at the average thickness $<d>\approx 4$ nm[8,10] and Ag films reach it at $<d>\approx 9.5$ nm[11]; Cu films do not reach the bulk conductivity in a thickness range where the film is IR transparent. Fig. 1 shows transmittance spectra measured during silver-film growth. The spectrum for $< d > = 9.5$ nm (at the bottom) can be calculated within experimental errors (about 10% in film thickness) with the bulk data[6,12]

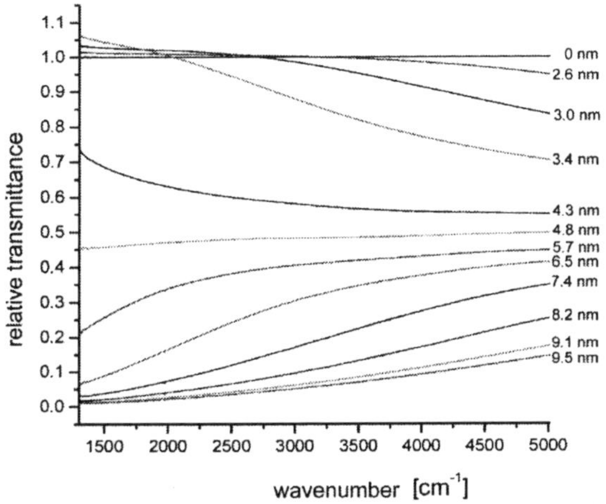

Fig. 1: Relative IR transmittance of silver films on UHV-cleaved MgO measured at normal incidence of light during film growth. The average film thickness $< d >$ as calculated from the deposition rate and deposition time is given. For experimental details we refer to Ref. 6, for example. Below 2000 cm^{-1}, where MgO phonon absorption starts, metal islands cause a reflection decrease of the MgO substrate.

of the dielectric function. In contrast to copper films grown under the same conditions[13], silver forms a smooth film with marginal amount of volume defects. Atomic-force pictures (immediately taken after the UHV studies) of the film with 9.5 nm in average thickness corroborate that result. They show a smooth film with shallow rectangular nanoholes that certainly are the result of strain release.[11,13] Furthermore, after CO exposure at 50 K, the film does not show any SEIRA signal of CO on island-like metal films.[13]

At the percolation threshold and below metal film spectra can be described with effective-media models, for example, if the film morphology meets the basic assumptions, with the two-dimensional Bruggeman approximation.[2] The percolation threshold is indicated by the nearly frequency-independent transmittance.[14] Deviations from that behavior arise then if the Drude parameters are strongly frequency dependent, as in the case of iron.[8] The film of Fig. 1 has its percolation threshold at about 4.5 nm of average thickness. The percolation threshold can be shifted towards lower thickness by several means: Gas exposure[15], additional nucleation centers on the substrate[16], lower substrate temperature, and higher deposition rate, respectively. The two last mentioned means lead to a decrease in crystalline quality of the film and, therefore, need post-annealing procedures.

Surface friction and charge transfer

Coverage changes down to the sub-monolayer range can be detected with IR transmittance spectroscopy of metal films. This concerns the metal-film growth itself[17] and also the broadband change due to adsorbates. Adsorbates cause changes of the dynamic conductivity via charge transfer that modifies the average plasma frequency of the film, via the surface-friction effect[18] that increases the scattering rate ω_τ, and in case of an island-type film, via the polarizability of the adsorbates between islands.[19] Eq. 3 roughly describes the relative adsorbate induced broadband transmittance change for a continuous film:

$$\frac{\Delta T}{T} = -\frac{2\frac{\omega}{c} d\left[\operatorname{Im}\varepsilon_{\text{film}} \cdot \left(\frac{\Delta\omega_p^2}{\omega_p^2} + \frac{\Delta\omega_\tau}{\omega_\tau}\left(\frac{\omega^2 - \omega_\tau^2}{\omega^2 + \omega_\tau^2}\right)\right)\right]}{\left(1 + n_{\text{substrate}} + 2\frac{\omega}{c} d \operatorname{Im}\varepsilon_{\text{film}}\right)} . \quad (3)$$

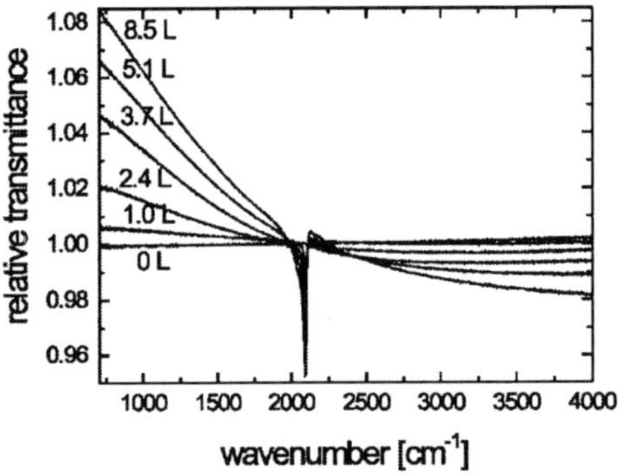

Fig. 2: Relative transmittance after CO exposure of 5 nm Cu (deposited onto KBr(001) at 100 K). 5 The exposures are given in the figure. The temperature of the experiment was 100 K.

The main advantage with respect to dc studies is the possible separation of changes $\Delta\omega_p^2$ of the squared plasma frequency and of changes $\Delta\omega_\tau$ of the scattering rate. Best fit results for CO on copper and iron films, for example, correspond to $\Delta\omega_p^2 > 0$, which is in accord with CO chemisorption models.[7,19] Adsorption of atomic oxygen gives $\Delta\omega_p^2 < 0$ indicating charge localization due to oxidation.[7] The change $\Delta\omega_\tau$ is caused by a dynamic effect where for a certain lifetime τ_{eh} conduction electrons from the metal are transferred to a molecular orbital leaving back a hole. After being transferred back the electrons have lost their phase coherence with the plasmon ensemble. This gives $\Delta\omega_\tau \propto 1/\tau_{eh}$. This electron-hole pair excitation is the more probable the closer that orbital to the Fermi energy.[18] Certain molecular vibrations that modulate the energetic position of the responsible orbital are essential. In other publications we have shown that this mechanism may give additional SERS and IR activity for molecules chemisorbed on atomically rough films.[20,21] Fig. 2 shows the broadband change due to CO adsorption on a Cu film that is not ideally smooth. Nevertheless, the two parameters $\Delta\omega_p^2$ and $\Delta\omega_\tau$ can be determined since the film is well above percolation, see Ref. 19.

Surface enhanced IR absorption (SEIRA)

In the spectra of Fig 2 the CO stretch mode is visible because vibration dipoles of molecules at sidewalls of Cu islands on KBr are along the field direction.[22] Their IR intensity appears to be enhanced and the line shape deviates from a Lorentzian.[2, 22] The effect is a Fano-type one where a single localized oscillator is coupled to a continuum of excitations. In this case the continuum is the plasmon-polariton spectrum of the metal-island film. The simplest description of this coupling is given by an effective-medium model for a mixture of the two types of polarizabilities. If the islands are small compared to the wavelength the result is maximum signal size from the adsorbate and maximum line

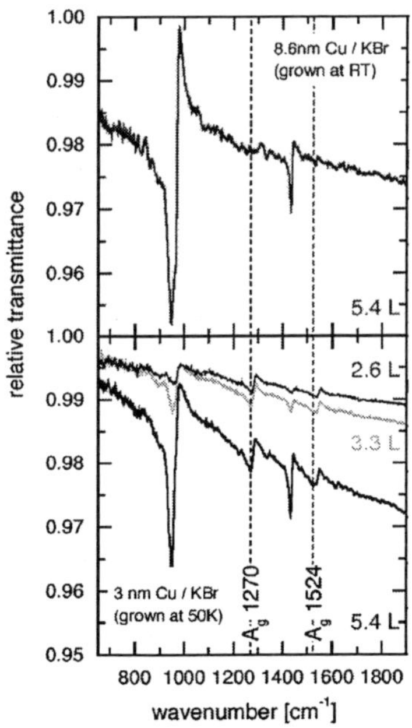

Fig.3: Relative transmittance of Cu films after C_2H_4 exposure given in Langmuir (L) at 50 K. The cold-condensed Cu film (bottom) shows features additionally to the lines of IR active modes.[21] Frequencies are given in wavenumber units (cm^{-1}).

assymmetry at percolation.[2] Important is the field enhancement between the metal islands, but for the signal shape, also phase relations between the complex interacting fields. In general, SEIRA is observed for adsorbates on nanostructures with complex dielectric function, i.e. metal structures or structures from ionic material in the range of phonon-polariton resonances. It occurs for instance on the various metal nanostructures (rough nanofilms, island films, grids, photonic arrays). The enhancement may reach three orders of magnitude[23] and is mainly due to the so-called field enhancement (in the nearfield of metal particles [2,3], at so-called hot spots[24] of disordered metal-island films, at plasmon resonances of periodic structures[25]). It can be further increased due to the "chemical effect" that usually means an increase in polarizability by adsorption and due to the special "first layer effect" that is the non-adiabatic interaction between vibrations and electron-hole-pair excitations[2,20,21], see above. SEIRA is complementary to surface enhanced Raman scattering and in some cases an even more successful analytical tool[26], other advantages of the rather non-destructive IR spectroscopy not mentioned. The aspect of accompaniment may be demonstrated here shortly with the C_2H_4 example. The inversion-symmetric molecule gives significant IR lines at its A_g Raman modes. The molecule is known to adsorb on copper with the C–C axis parallel to the surface and with negligibly small geometric changes. Certain molecular deformations are only a slight C–C length change, slight bending of CH_2 away from the surface.[27] Thus, symmetry breaking cannot explain the remarkable IR intensity of the Raman lines. Also, the effect is observed only if the metal film for adsorption has roughness on the atomic scale that usually is formed at deposition onto very cold substrates.[2,20,22] Fig. 3 shows SEIRA spectra of C_2H_4 on two differently grown Cu-island films. Both the films are close to percolation which is at about 3 nm for deposition at 50 K and at about 8.5 nm for deposition at room temperature (RT), respectively, on KBr(001).[21] Interestingly, a similar effect is not observed for C_2H_6 adsorption[21] since it does not offer a suitable orbital close enough to the Fermi energy. As in the surface-resistivity effect that kind of IR activation of Raman modes is supposed to be due to dynamic charge transfer.[4,21]

Summary

The development of the metal-film conductivity with film growth can be fast and nondestructively observed with IRS. Qualitative information on morphology can be obtained easily. The film conductivity can be determined quantitatively without electric contacts. Adsorbates induce a broadband spectral change that yields both, the plasma-frequency change and the scattering-rate change, *i.e.* charge transfer effects and surface resistivity contribution can be determined separately (for films beyond percolation). An island-like morphology leads to even bigger adsorbate induced spectral changes, in particular to SEIRA where asymmetric lines may appear. This asymmetry is maximum near percolation. Additionally, simultaneous excitation of adsorbate vibrations and electron-hole pairs may cause the occurrence of Raman lines in IR spectra of adsorbates in the first layer on metal surfaces with roughness on the atomic scale.

Acknowledgement

The authors (A. P. and A. P.) gratefully acknowledge financial support by the Deutsche Forschungsgemeinschaft (DFG).

References

[1] *E.g.* M. Osawa, in: J.M. Chalmers, P.R. Griffiths (Eds.), Handbook of Vibrational Spectroscopy, vol. 1, Wiley, Chichester, 2002, p. 785.
[2] A. Priebe, M. Sinther, G. Fahsold, and A. Pucci, J. of Chem. Phys. 119 (2003) 4887.
[3] A. Priebe, G. Fahsold, and A. Pucci, J. Phys. Chem. B 108 (2004) 18178.
[4] *E. g.* C. Siemes, A. Bruckbauer, A. Goussev, A. Otto, M. Sinther, and A. Pucci, J. Raman Spectroscopy 32 (2001) 231.
[5] M. Lust, thesis, Heidelberg University 2004.
[6] M.A. Ordal, R.J. Bell, R.W. Alexander, Jr., L.L. Long, and M.R. Querry, Appl. Opt. 24 (1985) 4493.
[7] G. Fahsold, M. Sinther, A. Priebe, S. Diez, and A. Pucci, Phys. Rev. B 65 (2002) 235408.
[8] G. Fahsold, A. Bartel, O. Krauth, N. Magg, and A. Pucci, Phys. Rev B 61, 14108 (2000).
[9] G. Fahsold, A. Priebe, N. Magg, and A. Pucci, Thin Solid Films 428 (2003) 107.
[10] G. Fahsold and A. Pucci, Advances in Solid State Physics (ed. by B. Kramer, Springer press) 39 (2003) 833.
[11] D. Seibel, diploma thesis, Heidelberg University 2004.
[12] H. J. Hagemann, W. Gudat, C. Kunz, J. Opt. Soc. Am. 65 (1975) 742.
[13] F. Meng, G. Fahsold, and A. Pucci, physica status solidi (c) in press (2005).
[14] *E. g.* S. Berthier and K. Driss-Khodja, Physica A 157 (1989) 356.
[15] M. Lust, A. Priebe, G. Fahsold, and A. Pucci, Surface and Interface Analysis 33 (2002), 487.
[16] G. Fahsold, A. Priebe, N. Magg, and A. Pucci, Thin Solid Films 364 (2000) 177.

[17] G. Fahsold, A. Bartel, O. Krauth, and A. Lehmann, Surface Science 433-435, 162 (1999).

[18] B. N. J. Persson and A.I. Volokitin, Surface Science 310 (1994) 314 .

[19] G. Fahsold, M. Sinther, A. Priebe, S. Diez, and A. Pucci, Phys. Rev B 70 (2004) 115406.

[20] M. Sinther, A. Pucci, A. Otto, A. Priebe, S. Diez, and G. Fahsold, physica status solidi (a) 188 (2001) 1471.

[21] A. Priebe, A. Pucci, and A. Otto, submitted to J. Phys. Chem. B (2005).

[22] A. Pucci, physica status solidi (b) 242 (2005) 2704.

[23] Y. Nishikawa, T. Nagasawa, K. Fujiwara, and M. Osawa, Vibrational Spectroscopy 6 (1993) 43.

[24] *E. g.* P. Etchegoin, L. F. Cohen, H. Hartigan, R. J. C. Brown, M. J. T. Milton, and J. C. Gallop, J. Chem. Phys. 119 (2003) 5281.

[25] Kenneth R. Rodriguez, Summit Shah, Shaun M. Williams, Shannon Teeters-Kennedy, and James V. Coe, J. Chem. Phys. 121 (2004) 8671.

[26] C. Domingo, J.V. García-Ramos, S. Sánchez-Cortés, J.A. Aznarez, J. of Molecular Structure 661-662 (2003) 419.

[27] H. Öström, A. Föhlisch, M. Nyberg, M. Weinelt, C. Heske, L.G.M. Pettersson, A. Nilsson, Surface Science 559 (2004) 85 and references therein.

AN AFM INVESTIGATION OF OLIGONUCLEOTIDES ANCHORED ON AN UNOXIDIZED CRYSTALLINE SILICON SURFACE

G. Longo[1], M. Girasole[1], A. Cricenti[1], F. Cattaruzza[2], A. Flamini[2] and T. Prosperi2

[1]CNR – Istituto di Struttura della Materia, Sezione di Tor Vergata Roma - Italy.
[2]CNR – Istituto di Struttura della Materia, Sezione di Montelibretti Roma - Italy.

Carboxylic terminated monolayers have been covalently attached on boron doped crystalline silicon (1,0,0) surfaces using a Cathodic Electro Grafting (CEG) technique. The functionalization concentration and efficiency have been evaluated through various surface analysis. In particular, topography images, performed with an Atomic Force Microscope (AFM), were used to optimise the protocol to obtain a most uniform monolayer.

Poli-thymine oligonucleotides (Poli-T) have been anchored on the functionalized surfaces to form a nano-bio sensing device, selectively reacting to a particular target molecule, the Poli-adenine (Poli-A). Characterization of these samples has been performed through surface imaging and quantitative fluorescence measurements. In particular, AFM images suggest that the DNA fragments self-assemble on the surface forming characteristic toroidal shapes

1. Introduction

Recently a great interest has arisen on the development of interfaces between molecular systems and traditional electronic or optoelectronic devices[1]. Such devices can have a great number of applications, with the only limitation residing in the ability to obtain reproducible and controllable preparations. Several other studies have concentrated their efforts on anchoring biomolecules to substrates as gold, metal oxides quartz, glass or silicon oxide [2-

6]. In this view we have optimized a technique to anchor oligonucleotides (used as active sensing components) on a crystalline (100) silicon surface which will allow a better and easier implementation of the biological components in the standard device manufacturing techniques.

Recently we functionalized a crystalline silicon surface with a long chain carboxylic acid terminated (10-undecynoic acid molecules) monolayer through stable covalent bonds.[7] We have shown that a Si-$(CH2)_n$COOH surface can be a good substrate to anchor amino terminated biomolecules such as enzymes, antibodies or oligonucleotides. Following these results we hereby show how oligonucleotides immobilized on such a surface can be a good sensing device. In particular we anchored a 20-Poli-T molecule on functionalized surfaces and we exposed such surfaces to a complementary 20-Poli-A oligonucleotide characterizing each step of this procedure with AFM and fluorescence spectroscopy.

In particular, the high resolution obtainable through the atomic force microscopy allows to detect small features and characteristics of the functionalization and linking procedures. As an example, the phase lag imaging of the functionalized silicon surfaces shows a structurization that can be associated to the different orientations of the molecules on the surface.

Topography images of the oligonucleotides anchored on the surface evidence characteristic toroidal structures that are formed by smaller granules.

2. Experimental Setup

Standard instrumentation has been used to characterize the samples. Fluorescence measurements were performed with a Perkin-Elmer MPF-44B spectrometer. The AFM images were taken with a home built microscope described in detail elsewhere [8] with a maximum scan range of 25 micron in x-y and 6 micron in z. Silicon Nitrade AFM tips were used. The measurements were performed in air, at room temperature and constant 30-35% relative humidity with a scan rate of approximately 0.1 lines/s. To obtain contact mode images, the microscope was operated in the weak repulsive regime (interaction between tip and sample <nN), allowing a high resolution imaging. For the non contact tapping mode images, the microscope was operated in the weak repulsive regime, coupling a very low mean tip-sample interaction (~pN) with a good lateral resolution and the ability to obtain phase lag images. No noise filter was applied to the raw data which was treated only by a background subtraction and, when necessary, a plane alignment.

2. Materials

2.1 Functionalisation of the Surface

A two step procedure is required to link a bio molecule on a silicon surface. At first a linker molecule layer must be placed to obtain a functionalized crystalline silicon surface only then the biomolecule can be anchored on this layer. Such a procedure has been described elsewhere using in particular gold or silicon oxide substrates. [2-6] These works also point out that the quality of the functionalized surfaces is a critical factor to obtain a good device.

Silicon boron doped wafers on the (1,0,0) orientation (Siltronix – France- resist. 0.1-0.01 Ωcm, 1" diam., 500 μm thickness) have been cleaned by means of a standard technique involving a fluoridric acid chemical etching to obtain a clean hydrogen-terminated silicon surface. This is well known to protect the surface from oxidation and is very reactive to form covalent bondings. [9]

To functionalize this surface a great attention was posed on the choice of an adequate cross linker molecule. This layer must allow the anchoring of biomolecules in biocompatible conditions. Furthermore, the linker molecule must keep the active element at an adequate distance from the silicon surface, to enhance it's sensing ability.

Most techniques and linker molecules are thoroughly studied in literature. We have chosen the 10-undecynoic acid molecule ($HC{\equiv}C\text{-}(CH2)_9COOH$) because of the very strong covalent bond that the triple C-C link forms with the silicon surface and because the terminal carboxylic group -COOH links with high efficiency with amminic groups –NH2. that are easily connected to biomolecules as lipids, oligonucleotides, etc. [10]

Linking a layer of 10-undecynoic acid on a surface can be performed by various methods. In particular three wet chemistry methods are well known: photo assisted, thermally assisted, and Cathodic (anodic) electro grafting. We have previously shown that a low current (<25 mC) cathodic electrografting deposition ensures a complete and uniform functionalization of the silicon surface [7].

The hydrolysis of the Si-H link with the C-C triple bond allows the formation of a covalent bond between the silicon and the carbon atoms (hydrosililation). These long molecules spontaneously pack on the surface maintaining their reciprocal positions through hydrophobic interactions. A thorough cleaning procedure removes the badly deposited or physisorbed molecules leaving a covalently bonded layer on the crystalline surface.

Since the carboxylic group must be chemically activated to allow the reaction with other molecules, this functionalization process protects the surface from oxidation or other chemical attacks [11,12].

2.2 Dna Deposition

The use of oligonucleotides as sensing devices has its origin in the great amount of information available on the interactions between different nucleotides. Specifically, an important feature is the very high selectivity in the interactions: a single strand nucleotide will link only with it's particular complementary nucleotide to form a double strand DNA fragment.

The molecule that has been chosen is a 3'-$(T)_{20}$-5' (20-fold thymine oligonucleotide) terminated with an amminic group available to react with the carboxylic terminal on the surface. For quantitative measurements a fluoresceine tag has been placed on the molecules.

As elsewhere described, to activate the cross linker on the surface, we work in a solution of EDC at pH 6.84. This solution allows the formation of a covalent bond between the –COOH and the $-NH_2$ groups while ensuring the DNA integrity [13].

The cleaning procedure, in this case, must remove all the molecules that didn't form a covalent link with the surface: when a single strand molecule lays on the surface for all its length or if a terminal group forms a much weaker saline bond. The fluoresceine tag allows to define when the cleaning procedure is complete: when the fluorescence of the cleaning residuals drops to zero, there are no mode oligonucleotides to be removed from the surface.

2.3 Hybridization

Every oligonucleotide-covered surface has been exposed to an adequate target complementary molecule. Since a thymine monomer reacts with a very high efficiency with an adenine monomer, the ideal target for these biosensors is a 5'-(A)20-3' (20-fold adenine oligonucleotide). As in the anchoring process a fluorescent tag is linked on every 3' terminal of the molecules. The exposure conditions must allow the hybridization of the two single strand molecules, leaving the Poli-T anchored on the surface. The exposure is performed at 90°C in a buffered Poli-A solution. Finally the temperature is gradually lowered to room temperature to stabilize the T-A link. This ensures the formation, on the silicon surface, of double strand oligomers with a double fluoresceine tag. The cleaning procedure is performed at 30°C using pure water and buffer solution. This ensures the removal of most of the physisorbed species (salt or un-hybridized molecules) while maintaining the reacted DNA fragments on the surface.

3. Results and Discussion

Different surface measurements were performed to characterize the steps of the sample preparation. In particular, AFM and fluorescence analysis has allowed to optimize every single step of the procedure and have evidenced interesting properties of the samples.

3.1 Fluorescence analysis of the Functionalized samples

To estimate the concentration of 10-undecynoic acid molecules present on the surface after the functionalization procedure, quantitative fluorescence measurements were preformed.

An AMC fluorophore has been substituted to the terminal –COOH group to tag the molecules on the surface. This procedure is well known in literature and has a ~60% efficiency [7]. After a thorough cleaning, heterogeneous hydrolysis (a standard technique with an efficiency above 90%) has been used to bring all the fluorescent AMC molecules in solution substituting them with a new –COOH group. This procedure, through a simple quantitative fluorescence measurement, allows to estimate the concentration of the functionalizing species on the surface in 0.44 ML [7].

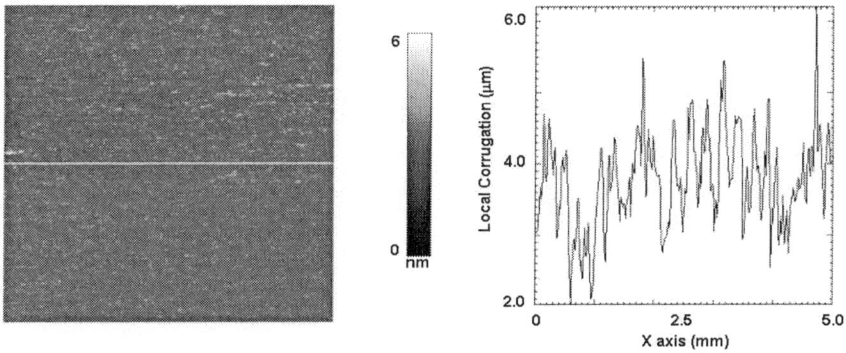

FIG. 1 5.0 μm x 5.0 μm contact mode AFM image of a functionalized sample. The surface appears uniform, as confirmed by the cross section, showing a local corrugation of 0.5-1.0 nm.

3.2 AFM analysis of the Functionalized samples

To characterize the quality of the functionalizing procedure, in particular the uniformity of the 10-undecynoic layer on the crystalline silicon surface, contact-mode AFM images were taken in different areas and on different samples as shown in fig 1. Cross section lines taken on these images show that the surfaces have a good uniformity. The good reproducibility on different samples ensures the quality of the preparation method.

The measurements on all these samples allow to calculate a local corugation of 0.5 nm, consistent with the presence of a single layer of 10-undecynoic molecules oriented perpendicularly to the silicon surface

3.3 Fluorescence analysis of the Single Strand samples

After the Poli-T anchoring on the surface, we must evaluate the efficiency of the process. Once again a quantitative fluorescence technique allows to estimate the oligonucleotide concentration on the surface. Using a high efficiency endonucleases, all the fluoresceine molecules present on the silicon surface are brought in solution and thus can be measured through a quantitative fluorescence technique. The maximum concentration we have achieved using our procedure is 2.44×10^{12} mol/cm^2. This value is perfectly comparable with the results present in literature regarding anchoring on similar kinds of substrates but is much lower if compared with the theoretical maximum that is 36×10^{12}

mol/cm^2.[14] As will be evident in the following paragraphs, a too high nucleotide concentration could jeopardize the sensing ability of the device.

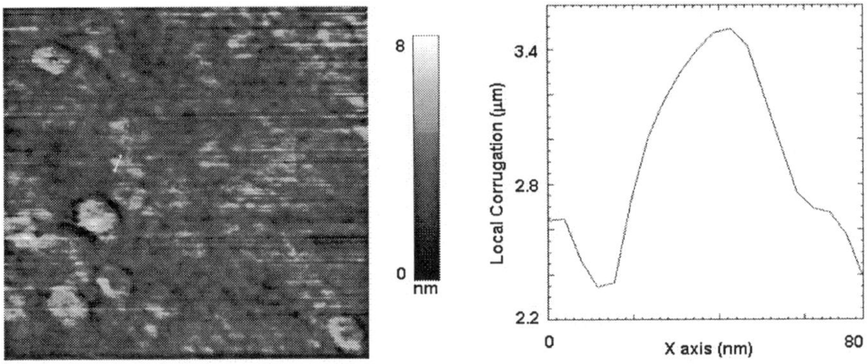

FIG. 2 2.5 μm x 2.5 μm tapping mode AFM topography image of a ssT_{20} covered sample. The molecules self assemble to form typical toroidal structures..

3.4 AFM analysis of the Single Strand samples

Organic molecules, and in particular DNA molecules, are a very reactant species. They can link to anything that comes to a too high interaction with the sample. Thus, using a standard contact mode AFM the risk of "cleaning up" the sample instead of measuring it exists. Consequently, an ideal technique to analyze the morphology of any ultra-soft sample, and in particular of oligonucleotides on a silicon surface, is an intermittent contact mode AFM (Tapping AFM) in which the tip oscillates on the sample and the mean interaction between the tip and the sample during a complete oscillation is of the order of the tenth of pico Newton while the maximum interaction is below the nano Newton level and is present only for a very small fraction of the time per oscillation.

With this technique, a high resolution surface characterization of the functionalized, oligonucleotide-bonded, crystalline silicon surface can be performed with a high resolution and very limited sample deterioration.

The image in fig 2 shows a quite homogeneous distribution of toroidal structures with a lateral dimension that ranges between 150 and 300 nm. These structures, as shown in the cross section, are formed by an aggregation of smaller 40 nm wide and 1.5-3 nm high structures. The toroidal structuring is a common feature of most of the samples analyzed.

3.5 *Fluorescence analysis of the Double Strand samples*

After exposing the oligonucleotide based sensor to the complementary molecule (Poli-A oligonucleotide), an hybridization efficiency measurement must be performed. The main difference with the single strand case is that the double strand oligonucleotides are linked to two fluoresceine molecules: one connected to the Poli-T and the other to the Poli-A. To estimate the hybridization efficiency we must compare the Poli-A concentration with the Poli-T concentration. We must bring in solution only the fluorophore linked to the Poli-A molecules while leaving the Poli-T firmly on the surface. This can be achieved considering that at 45°C the adenine-thymine link undergoes a melting procedure and breaks spontaneously. So, bringing the samples to 90°C in a 7M urea solution will leave a single strand Poli-T molecule layer on the surface (which is unscathed by the temperature and the solution) and all the Poli-A molecules in solution. The fluorescence concentration thus obtained must be compared with the Poli-T concentration on the surface. This can be obtained exposing the sample to endonucleases as described earlier.

This measurement method allows to estimate that the full hybridization of all the molecules on the surface occurs if the Poli-T concentration is 2.2×10^{12} mol/cm^2. This is very close to the maximum concentration of Poli T we obtained. A higher value would pack the molecules on the surface limiting their ability to interact with the target species.

3.6 *AFM analysis of the Double Strand samples*

As for the single strand covered samples, the morphological imaging has been performed on these samples using a Tapping mode AFM. As shown in Fig 3, the surfaces show an uniform distribution of toroidal structures similar to those already seen on the single strand covered surfaces. Though the form of these structures seems similar to the single strand based ones, the dimensions are significantly different. The lateral dimensions are ranging between 100 and 200 nm while the heights vary between 2 and 4 nm. Thus, the double strand covered surfaces confirm the spontaneous ordering described for the single strand covered samples. The larger observed height of the structures brings about the mechanism of AFM imaging and can be justified by the greater stiffness of a double stranded DNA fragment compared to a single strand DNA. Thus, the interaction with an AFM tip will produce a smaller deformation of the molecule, resulting in a higher and slimmer structure.

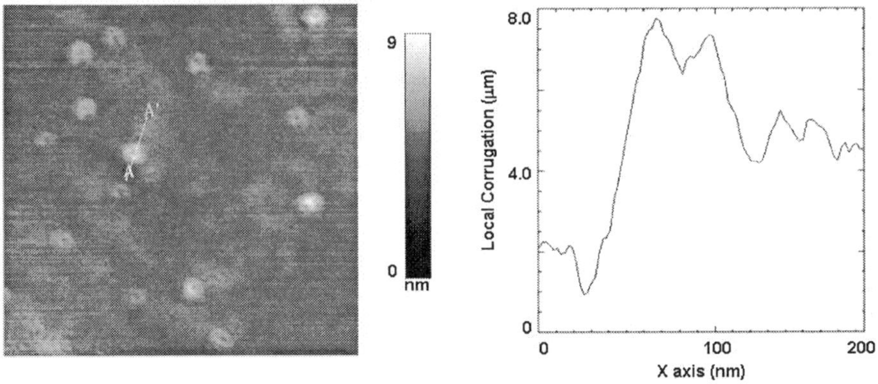

FIG. 3 0.6 µm x 0.6 µm tapping mode AFM topography image of a $dsTA_{20}$ covered sample. Also in this case, the molecules self-assemble forming typical toroidal structures.

The double strand covered surfaces confirm the spontaneous ordering described for the single strand samples.

4. Conclusions

Following our previous work on the silicon functionalization techniques, we have covered crystalline, boron doped, (100) silicon samples with a carboxylic acid terminated monolayer. This functionalization has been exploited to anchor an amino-terminated oligonucleotide to obtain a working biosensor. This has been performed through exposition to the complementary oligonucleotide, obtaining a double strand covered silicon surface. Each phase of the preparation has been characterized with AFM surface analysis and quantitative fluorescence based efficiency measurements. Thus we have verified that such a procedure allows the formation of an active oligonucleotide layer on the functionalized surface, reacting to a specific oligonucleotide.

The quantitative fluorescence measurements on the oligonucleotide covered samples evidence that this anchoring process has a concentration limit around $2.5x10^{12}$ molecules/cm^2. Upon exposure to the target molecule, the quantitative fluorescence measurements allow to estimate the reaction efficiency. A full hybridization (100% efficiency) can be obtained with a single strand concentration of $2.2x10^{12}$ molecules/cm^2, since a higher concentration would suffocate the sensing ability of the molecules.

Finally, AFM imaging on single strand and double strand covered silicon surfaces evidence the spontaneous formation of typical toroidal forms, formed by the aggregation of four or more smaller structures. The dimensions of these smaller components are compatible with the dimensions of oligonucleotide molecules, confirming that the toroidal forms probably originate from a self-assembly process during sample preparation.

REFERENCES

1. P. Fortina, L.J. Kricka, S. Surrey and P. Grodzinski, *Trends in Biotechnology*, **23**, 2005, 168
2. T.M. Herne and M.J. Tarlov, *J. Am. Chem. Soc.*, **119**, 1997, 8916
3. T. Vo-Dinh, J.P. Alarie, N. Isola, D. Landis, A.L. Wintenberg and M.N. Ericson, *Anal. Chem.*, **71**, 1999, 358
4. J.B. Lamture, K.L. Beattie, B.E. Burke, M.D. Eggers, D.J. Ehrlich, R. Fowler, M.A. Hollis, B.B. Kosicki, R.K. Reich, S.R. Smith, R.S. Varma and M.E. Hoga, *Nucleic Acids Res.*, **22**, 1994, 2121
5. J. Xu, J.J. Zhu, Q. Huang and H.Y. Chen, *Electrochemistry Comm.*, **3**, 2001, 665
6. R. Lenigk, M. Carles, N.Y. Ip and N.J. Sucher, *Sensors and Actuators B*, **68**, 2001, 100
7. F. Cattaruzza, A. Cricenti, A. Flamini, M. Girasole, G. Longo, A. Mezzi and T. Prosperi, *J. of Mat. Chem*, **14**, 2004, 1461
8. A. Cricenti and R. Generosi, Rev. Sci. Instrum. **66**, 1995, 2843.
9. D.D.M. Wayner and R.A. Wolvow, *J. Chem. Soc.*, **2**, 2002, 23.
10. L.F. Fieser and M. Fieser, *Reagents for Organic Synthesis*, vol. 1,Wiley, New York, 1967, p. 364
11. E.G. Robin, M.P. Stewart and J. Buriak, *Chem. Commun.*, 1999, 2479
12. C. Gurtner, A.W. Wun and M.J. Sailor, Angew. *Chem., Int. Ed.*, **38**, 1999, 1966–1968.
13. F. Wei, B. Sun, Y. Guo and X.S. Zhao, *Biosensors and Bioelectronics*, **18,** 2003, 1157
14. S.O. Kelley, N.M. Jackson, M.G. Hill and J.K. Barton, Angew. *Chem. Int. Ed. Engl.*, **38**, 1999, 941.

A NEW APPROACH TO CHARACTERIZE THE POLYMERIC NANOFILTERS CONTAMINATION LEVEL USING SCANNING NEAR-FIELD OPTICAL MICROSCOPY

C. OLIVA[*], A.USTIONE[*], A. CRICENTI[*,†], V. CECCONI*, E. CURCIO[‡]

[*]*Istituto di Struttura della Materia CNR, Via del Fosso del Cavaliere, 100, Rome, Italy*

[†]*Dep. of Physics and Astronomy, Vanderbilt University, Nashville, TN 37212 USA*

[‡]*Institute on Membrane Technology CNR-ITM, Via P. Bucci, 17/C, Rende (CS) Italy*

Synthetic polymers such as polysulfone, polyethersulfone, polycarbonate, polyamide cellulose acetate, are widely used as filtration membranes in different fields as pharmaceutical, agro-industrial, biotechnological ones and in the water treatment. The use of Atomic Force Microscope (AFM) has received considerable attention within the membrane research community, as a tool to investigate main membrane properties. This is due to the fact that it has the capability to examine morphology over a wide range of dimensions, from microns down to the molecular level and allows imaging even on soft surfaces both in air and liquid environments.

We propose the Scanning Near-Field Optical Microscope (SNOM) as a new approach, able to investigate not only the membrane morphology, but also the correlation between surface topography and chemical properties at a nanoscopic scale. Even if we are still at a preliminary stage, the aim of our work is to set-up a reliable procedure to characterize nanofilters and quantify their rejection capacity. SNOM gives us the possibility to identify the different contamination agents, also in hydrophilic membranes, working with a wide spectral range of wavelengths.

The information obtained from the studies conducted by scanning probe tool, can be used either directly or indirectly to develop predictive modeling for nanofiltration membrane system.

Nanofiltration[1,2] is the most recently developed pressure-driven membrane process for liquid-phase separation, it is used when low molecular weight solute have to be separated from solvent. Compared to ultrafiltration (UF), nanofiltration (NF) membranes have a smaller pore size so that the advantages of NF process are: a low operating pressure[3] compared to reverse osmosis (RO) (when the transport is governed by a diffusion mechanism) and a high rejection of organics compared to UF. The permeate quality depends largely on the type of membrane used.

The events controlling the performance of a nanofiltration membrane take place at a length scale of the order of few nanometers[4], some membrane parameters are particularly important, among them: pore distribution, scaling, membrane morphology and fouling tendency.

Dramatic declines in system flux and solute rejection are often associated with the occurrence of scaling and membrane bio-fouling. Scaling means the deposition of particles on a membrane and this process causes it to plug and the nominal flux to decrease. Bio-fouling contamination occurs most often during nanofiltration, it is a major impediment to the nanofiltration process efficiency. It can occur rapidly and can be ascribed to the complex growth of bacteria.

It would be very important to improve the performance of nanofiltration process through the systematic study about the types of membrane contamination. The bacteria growth factors and their concentration on a membrane system depend on critical factors such as temperature, the presence of sunlight, pH, dissolved oxygen concentration, and the presence of organic and inorganic nutrients.

Atomic Force Microscopy (AFM) is a valuable technique to investigate and quantify key properties of such membranes[5,6,7], both their fundamental form and their fouling properties. The technique has many advantages, the resolution is high and no damage on the sample occurs.

The morphology of the membranes surfaces was studied by AFM in terms of surface corrugation. The corrugation is defined as the difference between the higher and the lower points in the whole image. In the present study the same tip was used for all measurements and all captured surfaces were treated in the same way.

The AFM studies were conducted on both tapping and contact mode. The membranes used are polymeric nanofilters, named polyethersulfone[8] (PES) individuated by the initials NFPES10.

The Fig.1 represents an example of AFM topographic image (40 μm × 40 μm) in contact mode. The cross section (figure on the right) shows a good spatial resolution and a high signal to noise ratio.

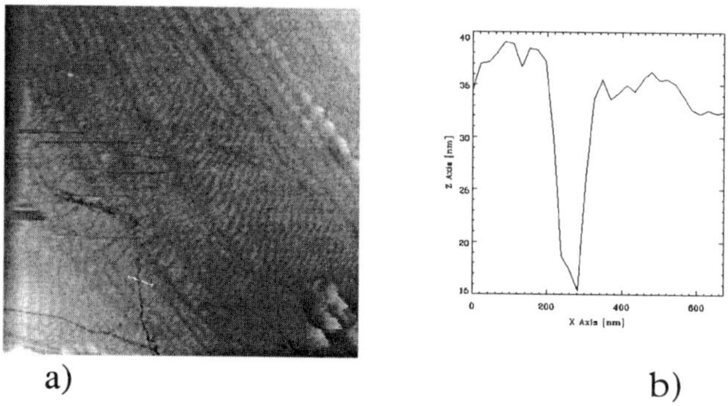

FIG. 1: a) 40 μm × 40 μm AFM-contact image on a long time air-exposed membrane. Although the good signal to noise ratio, b), the macrostructures on the surface due to the high hydrophily obscure the nanoscopic properties. It is required to treat the membrane surface, before AFM imaging, to quantify the extent of pollution just observing the morphology.

AFM software[9] was used for analysis of images. The color intensity shows the vertical profile of the sample with the light regions being the highest points and the dark regions being the depressions.

The membrane morphology has been studied with the Scanning Near-Field Optical Microscope (SNOM) also, and the same software for image analysis was used. An example (40 μm × 40 μm) of the recorded images is shown in the Fig.2 together with the cross section which suggests us the lateral resolution of the tip.

The images obtained with both the AFM and the SNOM, indicate a high membrane hydrophily and highlight that the water excess on the surface prevents us to appreciate the details and to give the quantitative interpretation of the parameters used to define surface features (peak-valley distance and mean pore diameter). The parameters were however roughly estimated for all studied surfaces; the values of the peak-valley distance are between a few tens and several hundreds of nanometers, it depends on the extent of hydrophily, and the values of the mean pore diameter are around 100 or 200 nanometers.

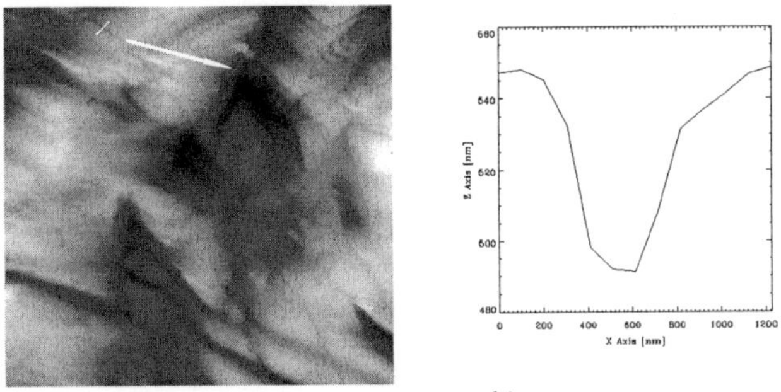

a) b)

FIG. 2: a) 40 μm × 40 μm SNOM image on an air-exposed membrane. Although the good signal to noise ratio, b), the macrostructures on the surface due to the high hydrophily, obscure the nanoscopic properties, like pores, which are just weakly visible on the left of the topographic image. Collecting optical images in a wide range of wavelengths, the SNOM system is able to investigate the chemical properties of the nanofilters and to distinguish the different kinds of bio-fouling agents. Our aim is to improve the system to obtain a complete characterization of the hydrophilic membranes at a nanoscopic scale.

To remove the water excess on the membrane surface, we have applied the water/ethanol exchange procedure10 and we have monitored the degree of 'cleanness' by FTIR HATR (Fourier Transform Infra Red Horizontal Attenuated Total Reflection) spectra, represented in Fig.3.

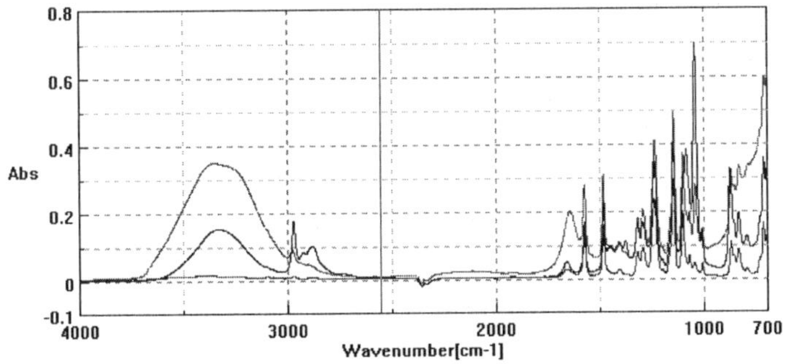

FIG. 3: FTIR spectra on a NF10PES membrane. The green line represents the nanofilter in presence of the water excess on the surface, the blue line is the HATR membrane spectra after water/ethanol exchange procedure, and the red one after a drying procedure.

This method has allowed us a chemical characterization of the unused-nanofilters and it can be used also to identify the absorption wavelengths of main contaminants, so that we can proceed chemically mapping the membranes with the SNOM system. So treated used and unused-nanofilters will be investigated exploiting the optical performance of the SNOM, particularly utilizing the versatility in working with light sources in a wide spectral range. The goal of this work is to obtain an efficient experimental method that allows to distinguish different contamination agents and to set a scale to evaluate the membrane bio-fouling level (compared to unused membranes). At the same time we aim to individuate the main parameters that determine the life-time of the nanofilters. We hope that information that can be obtained from these studies in conjunction with mathematical models, will be useful to give an improving in the membrane production processes.

REFERENCES

[1] A. H. Bannoud, Desalination 137 (2001) 133-139

[2] J. Schap, B. Van der Bruggen, S. Uytterhoeven, R. Coux, V. Vandecasteele, D. Wilms, E. Van Houtte, F. Vanlerberghr, Desalination 119 (1998) 295-302

[3] R. A. Bergman, Journ. AWWA 88 (1996) 32-43

[4] W. R. Bowen, J. S. Welfoot, Desalination, 147 (2002) 197-203

[5] A. Gordano, V. Arcella, E. Drioli Desalination accepted

[6] M. Hayama, F. Kohoroi, K. Sakai, J. Membr. Sci. 197 (2002) 243-249

[7]J. Y. Kim, H. K. Lee, S. C: Kim, J. Membr. Sci.163 (1999) 159-166

[8] http://www.mitsui-chem.co.jp/info/pes_e

[9] C. Barchesi, A. Cricenti, R. Generosi, C. Giammichele, M. Luce, M. Rinaldi, Rev. Sci. Instrum. 68 (1997) 3799

[10] S. Belfer, R. Fainchtain, Y. Purinson, O. Kedem, J. Membr. Sci. 172 (2000) 113-124

Magnetization reversal processes in Fe/NiO/Fe(001) trilayers studied by means of Magneto-Optical Kerr Effect

P. BIAGIONI

INFM - Dipartimento di Fisica, Politecnico di Milano,
Piazza Leonardo da Vinci 32, 20133, Milano, Italy
E-mail: paolo.biagioni@polimi.it

The magnetic properties of epitaxially grown Fe/NiO/Fe(001) trilayers have been investigated, for different thicknesses of the NiO spacer, by means of Magneto-Optical Kerr Effect. It was found that the NiO thickness $t_{\rm AFM}$ has a critical value $t_{\rm C}$ for the magnetic coupling between the Fe layers: for $t_{\rm AFM} < t_{\rm C}$ the magnetization directions align perpendicularly, in zero applied field, while the alignment is collinear for thicker spacers. A phenomenological model has been developed to reproduce and discuss the results.

1. Introduction

The coupling between ferromagnetic (FM) layers separated by a nonferromagnetic spacer has attracted many efforts in the last years, both for fundamental reasons[1,2,3] and for the possibility to develop applications such as magnetic sensors or magnetic recording devices.[4] The case of trilayers where two FM films are separated by a thin insulating antiferromagnetic (AFM) spacer can be of particular interest, because the indirect interaction due to conduction electrons is ruled out, and one would expect a large contribution coming from direct nearest-neighbour exchange across the spacer and the interfaces.

Here I report measurements of the magnetic properties of Fe(001)/NiO(001)/Fe(001) trilayers. The magnetization reversal processes have been studied by means of Magneto-Optical Kerr Effect (MOKE)[7] as a function of the NiO thickness $t_{\rm AFM}$. A simple numerical model has been developed in order to simulate the magnetization reversal and get a better insight in the physics involved.

2. Sample preparation

Fe(7 nm)/NiO($t_{\rm AFM}$)/Fe(300 nm) trilayers have been epitaxially grown on MgO(001) substrates. Samples were prepared in an Ultra High Vacuum chamber (5×10^{-11} Torr base pressure) by Molecular Beam Epitaxy on a MgO(001) single crystal substrate.[5] The first Fe layer was exposed to oxygen after growth, in order to obtain the very stable and well characterized Fe(001)$-$p(1×1)O surface[6] and thus lower the possibility of uncontrolled oxidation during NiO deposition. NiO was grown by evaporating metallic Ni in an oxygen atmosphere of partial pressure $p_{\rm O_2} = 7.5 \cdot 10^{-7}$ Torr. Cleanliness and surface reconstruction have been checked for each layer by means of X-ray Photoemission Spectroscopy and Low Energy Electron Diffraction, and finally all samples were capped with about 2 nm of Au to allow *ex situ* measurements.

3. MOKE results and numerical simulations

MOKE measurements were performed to obtain hysteresis loops of the samples and thus characterize their magnetic behaviour. Due to the penetration depth of the photons and the low thickness of the Fe overlayer and NiO spacer, the magnetization reversal behaviour of both FM layers is probed. MOKE measurements were performed in air and at room temperature, with the external field **H** applied along one of the in-plane easy axis of the Fe layers. Two distinct magnetization reversal behaviours are observed at different interlayer thicknesses, with a critical thickness $t_{\rm C}$ of about 4 nm for the transition between the two regimes. Below this value the relative alignment between the magnetization of the iron layers in zero applied field is perpendicular, and above it switches to collinear. No exchange bias has been detected, as expected in low-anisotropy NiO-based systems.[8]

In the left panel of Fig. 1 I show the experimental hysteresis loops for the Fe(7 nm)/NiO(1.4 nm)/Fe(300 nm) trilayer, i.e. for $t_{\rm AFM} < t_{\rm C}$. Both the in-plane longitudinal (parallel to **H**, $M_{||}$) and transverse (perpendicular to **H**, $M_{\perp}$) loops are shown. Also out-of-plane measurements have been performed, and no out-of-plane component of the magnetization vector has ever been observed. In the longitudinal loop, the inner narrow loop, with a coercivity of about 5 Oe, can be attributed to the Fe substrate, while the more complex behaviour of the outer loops, with the two lateral steps and a coercivity of about 300 Oe, comes from the contribution of the thin Fe overlayer. For the latter, at low applied fields, the overlayer magnetization is oriented perpendicularly to the substrate, as confirmed by the transverse ($M_\perp$) hysteresis loop, which shows a non-zero transverse component of the in-plane magnetization vector at correspondence with the intermediate plateaux of the longitudinal loop: they represent a possible intermediate state where the spins are directed perpendicularly to their initial and final directions, as a consequence of the fourfold symmetry of the Fe(001) overlayer.[9]

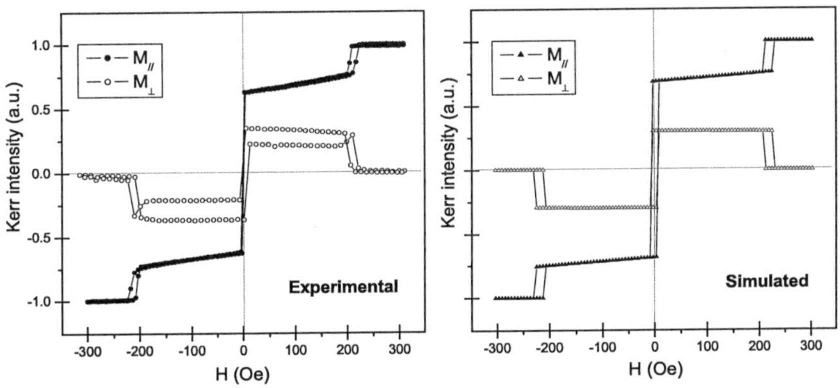

Figure 1. Experimental and simulated hysteresis loops for the Fe(7 nm)/NiO(1.4 nm)/Fe(300 nm) trilayer.

In the numerical simulation, the coupling between the two FM layers is described with a bilinear term, favouring collinear alignment, and a biquadratic term, favouring perpendicular alignment.[3] The magnetization of the substrate, much thicker than the other Fe layer, is supposed to follow the applied field, and then drive the magnetization reversal behaviour of the overlayer by means of the two coupling terms. Energy barriers for the nucleation of 180° domains are considered both in the iron substrate and overlayer. For the latter, the formation of 90° domains is also allowed, as confirmed by the experimental data. The values for all the energy barriers are taken directly from the experimental loops. The hypothesis underlying the simulation is that the sample can always be considered to be in a single domain state, as confirmed by the sharp transitions in the observed hysteresis loops.[9]

The expressions for the Fe substrate and Fe overlayer total energies per unit surface are the following:

$$E_{sub} = -HM_S t_{\rm sub}\cos(\vartheta - \varphi_1) + \frac{1}{4}K_4 t_{\rm sub}\sin^2(2\varphi_1),$$

$$E_{over} = -HM_S t_{\rm over}\cos(\vartheta - \varphi_2) + \frac{1}{4}K_4 t_{\rm over}\sin^2(2\varphi_2) + C_1\cos(\varphi_2 - \varphi_1) + C_2\cos^2(\varphi_2 - \varphi_1),$$

where H is the magnitude of the applied field, M_S the saturation magnetization (taken to be 1742 emu/cm^3, as in bulk Fe), ϑ is the angle between $\mathbf{H}$ and the [100] axis, $t_{\rm sub}$ and $t_{\rm over}$ are the Fe substrate and overlayer thicknesses, φ_1 and φ_2 are the angles between the magnetization directions and the same easy axis in the Fe substrate and overlayer, respectively. The cubic anisotropy constant K_4 has been experimentally measured to be $4.8 \cdot 10^5$ erg/cm^3, as in bulk Fe, while the coupling coefficients C_1 and C_2 are fitted to the experimental data.

The best fit for the Fe(7 nm)/NiO(1.4 nm)/Fe(300 nm) trilayer is shown in the right panel of Fig. 1, and is obtained with $C_1 = 4.35 \cdot 10^{-2}$ erg/cm^2 and $C_2 = 2.05 \cdot 10^{-1}$ erg/cm^2. There is actually a difference in the height of the plateaux in the transverse loop between the experimental data and our simulation: this behaviour can tentatively be attributed to the possibility that the Fe overlayer is broken into several 90° domains, rather than into a single one. This fact is not accounted for by the coherent rotation model.

Let's now turn our attention to spacer thicknesses above t_C. The hysteresis loop for the Fe(7 nm)/NiO(10 nm)/Fe(300 nm) trilayer is shown in Fig. 2. For this trilayer no significant transverse component of the magnetization vector has been detected. The intermediate plateaux in the loop reveal therefore a situation where the magnetization vectors of the two layers are antiparallel, as they switch at different coercive fields. An increase in spacer thickness decreases the value of the coupling constant for perpendicular alignment between the magnetic moments. The experimental data are indeed fitted with $C_1 = 4.35 \cdot 10^{-2}$ erg/cm^2 and $C_2 = 0$.

4. Discussions and conclusions

The value found for the critical thickness deserves some comments. In previous theoretical simulations,[2] the critical thickness for the transition between orthogonal and collinear coupling at zero applied field was calculated to be of the order of δ_W, where δ_W is the width of magnetic domain walls in the AFM spacer. This width has recently been experimentally measured, and reported to vary between 134 and 184 nm in NiO.[10] Our experimental

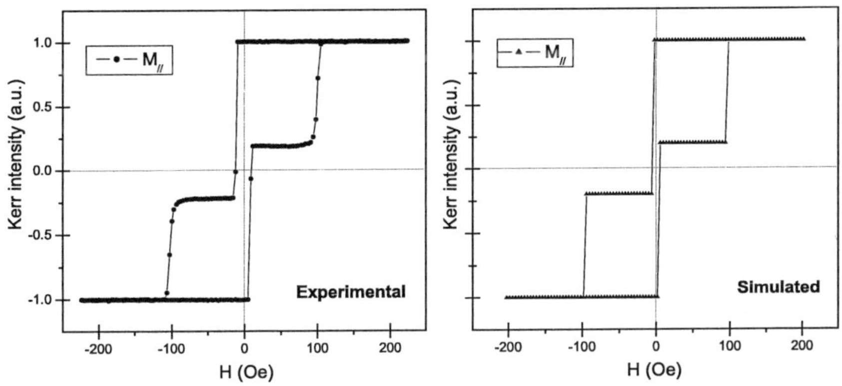

Figure 2. Experimental and simulated hysteresis loops for the Fe(7 nm)/NiO(10 nm)/Fe(300 nm) trilayer.

value for the transition between collinear and orthogonal coupling is quite smaller. A partial explanation could be that coherent rotation models[2] do not take into account the very common occurence of volume defects - such as vacancies or dislocations - in thin AFM films. These defects might reduce the characteristic length of the coupling effect in the sample and therefore account for our experimental value. It is also worth remarking that a similar critical thickness for the magnetic coupling between two FM layers separated by an AFM spacer has already been reported.[11,12]

It is also worth spending some more words on the two coupling terms used in the numerical simulation. The insulating character of NiO allows to exclude any oscillating behaviour of the bilinear coefficient C_1 due to the RKKY interaction. The phenomenological coupling terms reflect therefore the interplay between the exchange coupling at the interfaces and the static and dinamic properties of the AFM spin structure across the NiO. Slonczewski[1] proposed that the coupling through an AFM spacer should be described in the frame of the so called proximity magnetism model by a coupling term of the form $C_1(\varphi_2-\varphi_1)^2+C_2(\varphi_2-\varphi_1-\pi)^2$. The main difference between this phenomenological description and the bilinear and biquadratic terms is the behaviour at high applied fields, as the magnetization reaches the saturation value only asymptotically for the Slonczewski term. I tried to use this term for the energy expressions, but couldn't reproduce the experimental data, which show a saturation magnetization for finite values of the applied field. I can tentatively suggest that the Slonczewski term is a good description of the energetics of the magnetization reversal only for strong interface coupling, compared to the volume anisotropy energy.

In conclusion, I have studied the magnetization reversal behaviour of Fe/NiO/Fe(001) trilayers and found that there is a critical thickness of the NiO spacer for the magnetic coupling between the two Fe layers. Below this value, the relative alignment between the magnetizations of the iron layers in zero applied field is perpendicular, while above this value it switches to collinear. This occurrence can be related to the active role of the antiferromagnetic spacer whose spin structure mediates the exchange coupling between the FM layers. Simulation based on a bilinear and a biquadratic coupling term agree very well with the observed hysteresis loops.

Acknowledgments

I gratefully acknowledge my collegues at the Surface Physics group of the Department of Physics at Politecnico di Milano (Italy) and P. Vavassori from Università di Ferrara (Italy) for helpful collaboration and fruitful discussions.

References

1. J. C. Slonczewski, J. Magn. Magn. Mat. **150**, 13 (1995).
2. H. Xi and R. M. White, Phys. Rev. B **62**, 3933 (2000).
3. S. O. Demokritov, J. Phys. D **31**, 925 (1998).
4. *See e.g.* P. Grünberg, Phys. Today (May 2001) 31-37. G. A. Prinz, J. Magn. Magn. Mat. **200**, 57 (1999).
5. L. Duò, M. Portalupi, M. Marcon, R. Bertacco and F. Ciccacci, Surf. Sci. **518**, 234 (2002).
6. R. Bertacco and F. Ciccacci, Phys. Rev. B **59**, 4207 (1999).
7. E. Puppin, P. Vavassori, L. Callegaro, Rev. Sci. Instr. **71**, 1752 (2000). P. Vavassori, Appl. Phys. Lett **77**, 1605 (2000).
8. J. Camarero, Y. Pennec, J. Vogel, M. Bonfim, S. Pizzini, M. Cartier, F. Ernult, F. Fettar and B. Dieny, Phys. Rev. B **64**, 172402 (2001).
9. R. P. Cowburn, S. J. Gray, J. Ferré, J. A. C. Bland and J. Miltat, J. Appl. Phys. **78**, 7210 (1995).
10. N. B. Weber, H. Ohldag, H. Gomonaj and F. U. Hillebrecht, Phys. Rev. Lett. **91**, 237205 (2003).
11. M. E. Filipkowski, J. J. Krebs, G. A. Prinz and C. J. Gutierrez, Phys. Rev. Lett. **75**, 1847 (1995)
12. P. A. A. van der Heijden, C. H. W. Swüste, W. J. M. de Jonge, J. M. Gaines, J. T. W. M. van Eemeren, and K. M. Schep, Phys. Rev. Lett. **82**, 1020 (1999).

Laser-induced band bending variation for ZnTe (110)1x1 surface

S.D. Thorpe[1], S. Colonna[1], F. Ronci[1] and A.Cricenti[1]
B.A.Orlowski[2], I.A. Kowalik[2], B.J. Kowalski[2]

[1]*Istituto di Struttura della Materia, CNR, via Fosso del Cavaliere 100, 00133 Roma, Italy*
[2]*Institute of Physics, PAS, Al. Lotnikow 32/46, 02 668 Warsaw, Poland*

Zinc Telluride is a semiconductor of the $A_{II}B_{VI}$ family which has stimulated interest because of its applications in electro-optics, acousto-optics, green laser generation, and also as substrate material for II-VI laser diodes for optical data processing.
The electronic, optical and morphological properties of zinc chalcogenides have been studied from both the theoretical and experimental point of view [1-7]. Among these studies, optical methods have proved to be very useful in the study of surfaces. However, in order to investigate the optical properties of surfaces it is necessary to consider that light penetrates the solid for a depth much larger than the *z*-extension of the surface. It is possible to estimate that the surface contribution to the reflectivity of a semiconductor is of the order of a few percent. The problem of bulk and surface effects discrimination in the reflectivity spectra requires a very accurate stability control of the optical system during the experiment. In this frame Surface Differential Reflectivity (SDR) has proved to be very effective in the surface study [8].

The study of the electronic band perturbation near the semiconductor surface is very important for microelectronics applications, for example in the metal-semiconductor interface formation (Schottky barrier) [13]. Defects and dopants (acceptors or donors) are always present at semiconductor surface and represent a perturbation of the local charge balance and consequently a natural band bending [13]. We can use the Franz-Keldish (FK) [10,11,12] effect to study the Fermi level movement as a function of metal evaporation. The FK effect caused the reflectivity variation induced by surface band bending. Pumping electrons, by a laser, from the valence to the conduction band it is a way to compensate the band bending (surface photovoltage) [15]. The SDR measurement, with and without laser, gives a measure of the band bending. First of all it is very important to verify the quality of the investigated surface in order to understand if we can see the FK effect in a cleaved ZnTe surface.

The ZnTe sample was grown by a modified Bridgeman method at the Institute of Physics, Polish Academy of Science in Warsaw [9]. The crystal was not doped and resulted p-type due to acceptor levels attributed to Te vacancies. The sample prepared for the experiment was oriented by X-ray diffraction method. Clean mirror like (110) ZnTe surface was obtained by Ultra High Vacuum (UHV) cleavage. Before reflectivity measurement a sharp 1x1 LEED pattern was observed. The optical experimental setup consists of an ion Argon laser (5500 Å Ion Laser Technology, laser radiation between 457 and 514 nm) used to give raise to a band bending with a consequent observation of the FK effect. The laser was incident on the sample with an angle of about 30^0 off the normal to the surface while its power (P) was varied in the range 100-500mW to investigate its influence on the FK line shapes.

The SDR method is used to observe the FK effect [8].

$$SDR = \frac{\Delta R}{R} = \frac{R_{laser}(P) - R_{nolaser}}{R_{nolaser}} \qquad (1)$$

The light emitted from a 150W lamp passes through a lens and is separated into two beams through a semitransparent CaF_2 beam splitter. One beam (I) is focused through another lens on the ZnTe surface in the UHV chamber (10^{-9}mbar base pressure) while the second beam (I_0) is focused onto an aluminium mirror. The light then enters an EG&G PARC half-meter monochromator with an optical multichannel analyzer (OMA) mounted on the exit slit. The intensities of the two reflected beams are measured by the OMA that is computer controlled through an IEEE 488 interface. The emission lines from a Hg lamp were used to calibrate the wavelength response of the spectrometer to within 1meV for emission near the band-gap energy of ZnTe. The data stored are the values of the ratio between the reflectivity of the cleaved surface and the reflectivity of the dummy sample. The reflectivity spectra were measured when the laser was off ($R_{nolaser}$) and during irradiation with laser ($R_{laser}(P)$).

Figure 1 shows the change of reflectivity upon irradiating the sample with laser power between 100 and 500 mW. It is worth to note that the shape of the FK oscillation does not change with laser power.

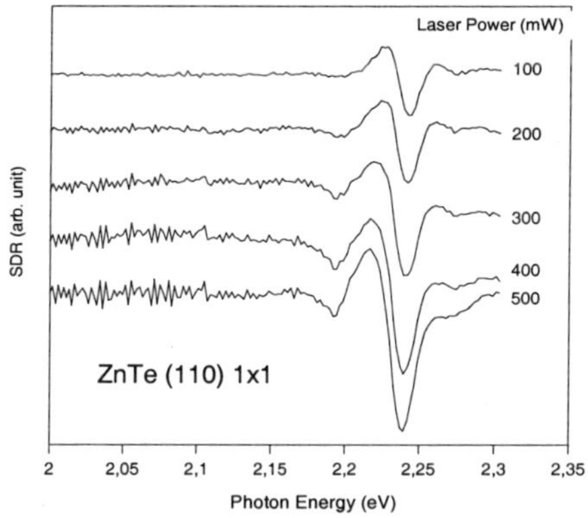

Fig.1.Reflectivity variation brought about by laser irradiation of the sample with laser power between 100 and 500 mW.

Figure 2 reports the amplitude of FK oscillation for increasing laser power values. This curves are similar to the analogous curves reported for CdTe [16].

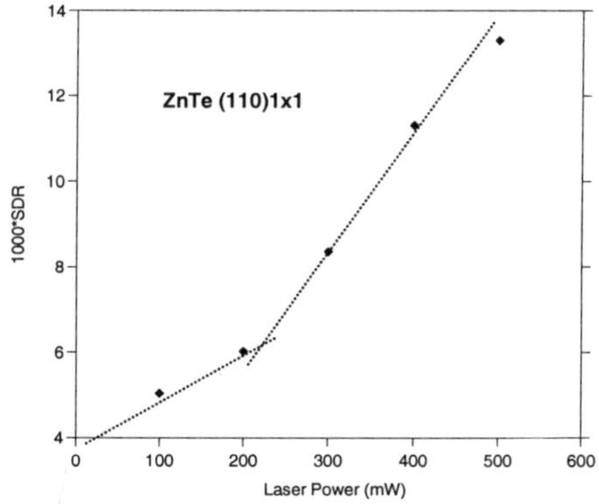

Fig.2.FK oscillation amplitude for laser power between 100 and 500 mW.

The increase of illumination intensity leads to the increase of the band bending and the Fermi level position changes to be located close to the middle of the band gap at the surface region. Strong uniform electric field influences the electronic states in the crystal. These electric field, created at the surface region, leads to the different wave function ψ of electrons in these region in comparison to the wave function ψ_o of it in the volume of the crystal. The momentum of electron is influenced by the term dependent on electric field. The problem is not stationary and cannot be solved by periodic Bloch function. It needs the application of time-dependent Schrödinger equation for a case of electron in uniform electric field and periodic electric field in the crystal. The problem was solved [12] for the case with assumption of small perturbation of the Bloch wave function at $t = 0$ by the electric field. As a finale solution it was found that due to electric field the step of absorption edge shifts to the lower energies (red shift). The value of the shift, in the frame of the assumptions of the solution [12] can be expressed by the equation:

$$\Delta\omega = \frac{AE^{2/3}}{m_{\|}^{1/3}} \qquad (2)$$

where A is a constant, E the electric field, $m_{\|}$ the effective mass of electron or hole on direction parallel to the direction of electric field. The shift $\Delta\omega$ increases with the strength of electric field and decreases with increase of carriers (electrons or holes) effective mass on the direction parallel to the electric field direction. Peak-to-peak distance on each curve increases with the increase of laser power radiation (see Fig. 2) and it corresponds to the increase of the FK effect. The measured amplitude of FK effect increases with the power of laser radiation as it is expected from formula (2). The inclination of the line depends on the $m_{\|}$ as an inclination coefficient of the dependence (2). This inclination coefficient changes with the power of laser radiation. It can be concluded that the change of inclination is caused by the appearance of dominant density of carriers with lower effective mass than it was for low power of laser radiation. For the case, in the region of low power of laser radiation the heavy carriers dominate while for high power of laser radiation the light carriers dominate. Sudden domination of new light carriers can appear due to the special band in the electronic structure of the crystal volume as well as due to the crystal surface effects. For low illumination we can expect domination of the high density of volume states in comparison to the surface. We expect that two inclinations of the curve presented in fig.2 corresponds to the two carriers of different effective mass. explanation of the origin of these heavy and light carriers needs some additional studies. A saturation of the FK signal is not observed up to the maximum laser power. This behaviour is different from the measurements reported in ref. [16] done on ZnTe(110) cleaved in air. It is possible to argue that the air exposure of ZnTe surface induces a partial saturation of the surface states responsible of the surface band bending.

Very sensitive SDR method was used to investigate the change of surface reflectivity caused by the variation of surface electric field created by sample illumination. We have measured the FK effect on clean ZnTe (110)1x1 surface as a function of laser power. The inclination of the linear increase in the FK amplitude (fig.2) has been observed to be lower in the range from 0 up to 250mW than in the range from 250 up to 500mW of laser power [16]. The energy resolution of FK method is much higher than of photoemission method (a few meV for FK against several tens of meV for photoemission). Using the FK effect no absolute value of band bending can be directly measured, but differences and variations in band bending may be determined with excellent precision and without disturbing the surface by the measurement itself [8].

Acknowledgement

This work was supported in part within MSRIT of Poland research projects 1 P03 B053 26 and 72/E-67/SPB/DESY/P-03/DWM 68/2004-2006 and PBZ/KBN/044/P03/2001.

References

[1] John P. Walter and Marvin L. Cohen, *Phys. Rev B* **1** 6 2661 (1970)
[2] Atsuko Ebina, Kiyomitsu Asano, and Tadashi Takahashi, *Phys. Rev B* **18** 8 4332 (1978)
[3] R. J. Meyer, C. B. Duke, A. Paton, E. So, J. L. Yeh, A. Kahn, and P. Mark, *Phys. Rev B* **22** 6 2875 (1980)
[4] James E. Bernard and Alex Zunger, *Phys. Rev B* **36** 6 3199 (1987)
[5] H. Qu, J. Kanski, and P. O. Nilsson, *Phys. Rev B* **43** 12 9843 (1991)
[6] A. Cricenti and B.A. Orlowski, *Phys. Rev B* **51** 4 2322 (1995)
[7] E. Erbarut, *Solid State Communication* **127** 7 515 (2003)
[8] A. Cricenti, *J. Phys.: Condens. Matter* **16** (2004) S4243-S4258
[9] Mycielski, A.; Szadkowski, A.; Lusakowska, E.; Kowalczyk, L.; Domagala, J.; Bak-Misiuk, J.; Wilamowski, Z. *Journal of Crystal Growth* **197** 423 (1999)
[10] W. Franz, *Z. Naturf. a* **13** 484 (1958)

[11] L.V. Keldish, *Sov. Phys.—JETP* **7** 788 (1958)
[12] L.V. Keldish, *Zh. Eksp. Teor. Fiz.* **34** 1138 (1958)
[13] H. Luth, *Surfaces and Interfaces of Solid Materials* (1995) (Berlin: Springer)
[14] B. O. Seraphin and N. Bottka, *Phys. Rev.* **145** 628 (1996)
[15] A. Bauer, M. Prietsch, S. Molodtsov, C. Laubschat, and G. Kaindl, *Phys. Rev. B* **44** 8 4002 (1991)
[16] B.A. Orlowski, I.A. Kovalik, B.J. Kowalski, M. Suffczynski, A. Mycielski, S. Colonna, C. Ottaviani, A. Cricenti, *J. Alloy and Compounds*, **382** (2004) 224

Optical Properties of Materials in an Undergraduate Physics Curriculum

Julio R. Blanco[(a)]
Department of Physics and Astronomy, California State University Northridge, 18111 Nordhoff Street, California, 91330-8268

Abstract

The need to introduce physics undergraduates to non-traditional subjects is ever increasing due to the job opportunities in interdisciplinary fields. The traditional upper-level curricula after the standard sequence in introductory calculus-based physics is challenging to many students. Adding more elective requirements is not in vogue with university administrators that must deal with a large influx of students with fewer resources. Experimental physics lends itself well to introduce students to interdisciplinary concepts. At California State University Northridge (CSUN), we have introduced modules in experimental physics to meet this need. All juniors and seniors are required to take two units of experimental physics per semester, a total of eight units. An experimental unit represents three contact hours per week. Each two units consist of two modules, each lasting seven and a half weeks, six hours per week. One of these modules exposes the students to thin film deposition by sputtering, imaging by scanning electron microscopy, and optical characterization using scanning ellipsometry. This early exposure to interdisciplinary applied physics motivates students and identifies difficulties with fundamental concepts.

Background

Many disciplines within the sciences are focusing in applied technologies and interdisciplinary fields. Chemistry departments are branching out into biochemistry. Biology departments are creating and expanding bioinformatics using protein, DNA and RNA sequencing data and adding biotechnology. Geology is branching out into environmental geology. Even the astronomy programs are moving away from direct observation and relying more on satellite astronomical data. This requires the students to learn about numerical analysis and programming to efficiently analyze the large data streams that are available. The mathematics programs are offering applied mathematics and statistical specialties designed for other fields such as business, law and others. The engineering programs are also defining more narrowly their degrees and adding more and more specialties.

In comparison, most physics departments have maintained the traditional approach. The traditional approach consists of preparing students to pursue a Ph.D. degree but not addressing the needs of industry or the interdisciplinary evolution that is taking place. This is not to say that physics departments are unaware of these changes. But it is more common to establish separate departments, such as applied physics, biophysics and materials physics, than to expand or transform traditional physics departments. As a consequence, the number of students majoring in traditional physics either does not grow or decreases.

The physics and astronomy (P&A) department at CSUN has added a bachelor of arts (B.A.) degree in Biomedical Physics. This degree offers a broad background in the sciences, including computer science, thus preparing the students for a wide variety of entry-level jobs in the marketplace and professional schools. In addition, the P&A department houses the Supramolecular Studies and the Computational Materials Theory centers as well as a joint Interdisciplinary Materials Center with the College of Engineering. These centers do provide opportunities for students that meet a certain level of academic background. Thus, the department is evolving, though slowly, with the needs of the students in their future careers.

The previous P&A physics program required that students take two or four experimental units. The department offered specialized labs in various disciplines. However, a student taking one of these courses will only be exposed to a very narrow field of traditional physics. The undergraduate program should offer the students a broad background in physics so that the students can better select the field of physics of interest. Certainly, the theoretical courses in the undergraduate program present the opportunity to learn the fundamentals of physics. When you add a couple of elective courses, it can be argued that the theoretical preparation is adequate. However, the experimental one is somewhat deficient.

The new P&A physics program requires that the students take eight units of experimental physics. Each unit can be in a different field of physics. Potentially exposing the students to eight different areas as opposed to one or two at best in the previous program. The concept is to take advantage of the expertise of the faculty members in the department. As the majority are involved in research, it is desirable to directly expose all the physics majors to these various research topics in a condensed manner.

A typical student enrolls in two units of experimental physics each semester during their junior and senior years. This represents six contact hours per week. We allocate seven and half weeks per module in a fifteen weeks semester. Each week, the students are responsible for six contact hours for a particular module. At the end of the seven and a half weeks, the student starts a different module. Alternatively, a student could take two modules simultaneously for fifteen weeks, dedicating three contact hours per week to each module. Either method, a student can complete two modules per semester.

A particular field or area of expertise can have more than one module, but each module should be independent of the others. We also include in our experimental modules getting proficiency in utilizing a high level technical computing language such as MatLab[1] that uses matrices and vectors for data visualization and analysis and numerical computation, and learning to use specialty software that teaches Monte Carlo simulations and similar applications in physics.

At CSUN, the P&A modules offer (or will soon offer) the physics majors opportunities in the areas of low energy nuclear physics, optics, microwaves, characterization of materials techniques, computational analysis in condensed matter, biophysics, general mathematics computational techniques, electron spin resonance and astrophysics. These modules can change over time to reflect the interests of the faculty members. Yet, the modules present to the students state-of-the-art experiences that hopefully help them make informed decisions as to their career future.

Optical Properties of Materials Module

One of the modules this author has introduced is to expose the students to the characterization of the physical properties of materials, particularly determining the optical properties. The module has been offered twice to date. Below a detailed description of the content of the course is described.

The students are introduced to sputtering using a small automatic sputter coater.(2) Thin films of gold are coated onto a variety of substrates such as a silicon single crystal wafer in any plane configuration, glass plates, metals and other materials as they are available. The sputtering conditions, such as deposition rates, pressure, ion current values and others, can be varied by small amounts resulting in different film surface finish as well as microstructure. The nominal thickness of the film can be found using a quartz crystal microbalance.

The students then observe their thin film samples using an optical microscope with 10-20 times magnification. Visual observations are made as to the quality of the surface of the film. The films are also observed using an electron scanning microscope (SEM) using 100-500 magnification.(3) Surface topography can be readily observed. The SEM also has an energy dispersive spectroscopy (EDS) device for composition analysis of the thin film sample using the INCA energy system.(4) The students learn qualitatively about the presence of impurities in their samples. Finally, the samples are measured using a spectroscopic ellipsometer in the range of 320-800 nm.(5) An analysis software is available for modeling the ellipsometric spectra providing the students with a basic understanding of thin film analysis.(6)

Each experimental technique used is introduced in a lecture type setting. The manuals for each instrument are readily available and the students are encouraged to familiarize themselves with the proper operation of each instrument. The instructor is available during the contact hours but the students have access to the instrumentation at other times. The most difficult instrument to use in this module is the SEM. A technician is available at all times that the student wishes to use this instrument. The student is allowed to manipulate the controls and obtain images of their samples. Reference samples are available that ensures the student is indeed observing the proper image of their sample.

The reflection equations of Fresnel are derived in the class. The case when there are two media separated by an ideal interface (no film) is considered. The case when there are three media: ambient, film and substrate is also discussed. The students learn how the equations are affected if the film is transparent (the extinction coefficient, $k = 0$) or not ($k \neq 0$). The students are asked to search the literature and find the reference data for those materials that are known.

Students can prepare a series of samples each of a different thickness. By modeling the ellipsometric spectra using regression analysis methods, the students are able to extract the film thickness for each sample. This can be compared with the nominal value selected during the deposition process. The optical constants (n and k) of the thin film can also be extracted and compared to the literature values.

The analysis software is simple to use. Once the students have learned the fundamental, they are able within a matter of a few hours, prepare, measure and analyze data. The results normally conform very well with expected outcomes. Even though the students are using sophisticated instrumentation and analytical tools, it is common for the students to comment how easy it is to carry out thin film analysis and to determine some optical properties of thin films. This can be taken as a measure of the success of the course. But they must be reminded that not every characterization they will encounter after school will be exactly the same.

Normally, the students are taken to other industrial and research laboratories to tour the facilities where these techniques are used. The students are expected to prepare a presentation that combines all their experiences in the course. All the students listen to each other's talks and are able to comment on each. Normally, a cohort of six students is typical during a module.

A typical presentation is twenty minutes in duration. The level of sophistication of these talks is very impressive. The students themselves are very critical of each other. It seems to this author that should the students wish to seek employment in this area, they will be able to represent themselves very favorably to a prospective employer after only a few weeks in this module.

In this module, the students are introduced to the fundamentals of thin film technology: the nature of physical sputtering, applications of sputtering to the deposition of thin films, film thickness and composition, direct observation of surfaces, polarized light, ellipsometry, optical properties of materials and layered structures, and optical data analysis.

At the start of the junior year the students have completed the standard three semester (twelve units) introductory fundamental physics series[(7)] that includes mechanics, electricity and magnetism, thermodynamics, optics, sound, waves and modern physics. The students also have completed sixteen units of calculus, including partial differential equations. The modules offered during the junior year cannot expect the students to be proficient above this level.

However, this author finds that the students can be introduced to the concepts of this module with little difficulty. In fact, it helps to better learn the concepts in the standard lecture courses that they must take during the junior year as part of the standard physics curriculum at CSUN.[(8)] A variety of books exist that can be used to supplement the course notes.[(9,10)]

Conclusions

The main objective for offering modules in experimental physics is to introduce the physics majors to a wide array of state-of-the-art instrumentation and computational techniques in a compact form without the need of advanced knowledge in the discipline. The trend to include undergraduates in research can be accomplished with this approach. Students early on in their career are made aware of the experimental nature of physics, its applications and its interaction with other disciplines. Certain difficulties that the students experience can be identified and corrective measures can be taken in the required courses in the major. This modular approach serves in a way to assess the success of the academic program by testing the students in their analytical and reasoning skills.

Acknowledgements

The author offers gratitude to David Carson for his assistance with the SEM and to Say-Peng Lim for his comments and suggestions. This work received partial support from NSF-EAR Grant No. 0216133, the Department of Physics and Astronomy, and the College of Science and Mathematics at California State University Northridge.

References

[a]Electronic mail: julio.blanco@csun.edu

[1]MatLab is supplied by MathWorks, Inc., 3 Apple Hill Drive, Natick, MA 01760-2098, USA. Telephone: 508 647 7000.

[2]The sputter coater is a Cressington 108 with an MTM-20 thickness monitor manufactured by Cressington Scientific Instruments, 34 Chalk Hill, Watford, WDI 4BX, England. Telephone: 1 923 220499.

[3]The SEM is a model QUANTA 600 manufactured by FEI Company, Podnikatelská 4, 612 00, Brno, Czech Republic. Telephone: 420 533 311 416.

[4]INCA Energy system is a high performance X-ray emission analysis tool manufactured by Oxford Instruments Analytical, Halifax Road, High Wycombe, Bucks, HP 12 3SE, England. Telephone: 44 (0)1494 479222.

[5]The spectroscopic ellipsometer is a model GESP5 manufactured by SOPRA, 26, rue Pierre Joigneaux, 92270, Bois Colombes, France. Telephone: 33 1 46 49 67 00.

[6]The software analytical tool is known as WINELLI and it is available from SOPRA.

[7]For example, Randall D. Knight, *Physics for scientists and engineers with modern physics: a strategic approach* (Pearson/Addison Wesley, San Francisco, California, USA, 2004). ISBN: 0-8053-8960-1, extended edition with MasteringPhysics.

[8]The junior standard curriculum includes: Mechanics, Introduction to Quantum Physics, Mathematical Physics I, Electricity and Magnetism, Foundations of Quantum Mechanics and an elective like Mathematical Physics II. Details about the physics program can be found at www.csun.edu/physics.

[9]Harland G. Tompkins and William A. McGahan, *Spectroscopic ellipsometry and reflectometry: a user's guide* (John Wiley & Sons, New York, 1999). ISBN: 0-471-18172-2.

[10]*Handbook of thin film technology*, edited by Leon I. Maissel and Reinhard Glang (McGraw-Hill, New York, 1983). ISBN: 0-07-039742-2.